A Plague of Frogs

A Plague of Frogs

UNRAVELING AN ENVIRONMENTAL MYSTERY

WILLIAM SOUDER

University of Minnesota Press
Minneapolis — London

Published by the University of Minnesota Press
111 Third Avenue South, Suite 290
Minneapolis, MN 55401-2520
http://www.upress.umn.edu

A Cataloging-in-Publication record for this book is available
from the Library of Congress.

Printed in the United States of America on acid-free paper

The University of Minnesota is an equal-opportunity educator and employer.

12 11 10 09 08 07 06   10 9 8 7 6 5 4 3 2

For Susan

Contents

Acknowledgments

MARY HADAR AND ROB STEIN, MY EDITORS AT THE *WASHINGTON Post*, took a chance on this strange story early and then stayed a long, sometimes winding course as it has played out. Their continuing encouragement and suggestions were invaluable. Many thanks to them both. Thank you, also, to Laurie Abkemeier, formerly of Hyperion, who was thinking the same thing I was thinking back in 1996 and helped make this book a reality. And thanks also to my agent, Elyse Cheney of Sanford J. Greenburger Associates, who got Laurie and me together. In addition to being a terrific agent, Elyse was a keen-eyed critic of the initial draft of the manuscript and provided many ideas for putting it right.

Deepest thanks to my editor at Hyperion, Mary Ellen O'Neill, for her support, clear thinking, and tirelessness in helping to untangle many a difficult passage, technical and otherwise. Mary Ellen's patience as the story continued to change in front of our eyes and her unwavering enthusiasm for the project saw me through to the end.

This book would not have been possible without the cooperation of the many people whose stories I have tried to tell. I am indebted to all of the scientists, landowners, teachers, and government officials who took time to talk with me. Among the many who helped were Cindy Reinitz and Jeff Fish; Don Ney; Dr. Shabeg Sandu, Dr. Gary Ankley, Dr. Sig Degitz, Dr. David Mount, Dr. Gil Veith, and Dr. Pat Schoff of the United States Environmental Protection Agency; Dr. George Lucier of the National Institute of Environmental Health Sciences; the Honorable Bruce Babbitt, Secretary of the Interior; Dr. Andrew Blaustein of Orgeon State University; Dr. Judy Helgen, Mark Gernes, Drew Catron, Jeff Canfield, Anna Bosch, Dorothy Bowers, Joel Chirhart, Joe Magner, and Ralph Pribble of the Minnesota Pollution Control Agency; Dr. Hillary Carpenter of the Minnesota

Department of Health; Gary Casper of the Milwaukee Public Museum and the Declining Amphibian Populations Task Force; Dr. David Wake of the University of California Berkeley; Dr. George Rabb of the Brookfield Zoo; Dr. Kathy Converse, Dr. Rebecca Cole, and Carol Meteyer of the National Wildlife Health Center; Laura Eaton-Poole of the United States Fish and Wildlife Service; Rick Levey of the Vermont Agency for Natural Resources; Jim Mumley of the J. M. Hazen Company; Dr. Robert McKinnell of the University of Minnesota; Erik and Larissa Mottl; Dr. Doug Fort of Stover Associates; Dr. Jim LaClair, of the Scripps Research Institute; Tom Meersman of the Minneapolis *Star Tribune*; Dr. Alan Pounds of the University of Miami; Jean Rodrigue of the Canadian Wildlife Service; Dr. David Green of McGill University; Dr. Dan Sutherland of the University of Wisconsin La Crosse; Dr. Mike Thurman of the United States Geological Survey.

Special thanks to the participants in the Oceanside Workshop: Dr. Val Lance, of the San Diego Zoo; Dr. Tyrone Hayes of the University of California Berkeley; Dr. Stephane Roy and Mark Carlson of the University of California Irvine.

Special thanks to Pieter Johnson and Kevin Lunde; Martin Ouellet of McGill University, Dr. Stan Sessions of Hartwick College, and Dr. Jim Burkhart of the National Institute of Environmental Health Sciences, all of whom spent many hours in conversation with me about their work and their thoughts on the deformities problem.

Dr. Sue Bryant and Dr. Bruce Blumberg of the University of California Irvine, and Dr. Ken Muneoka of Tulane University contributed enormously to my understanding of many highly technical issues. For that, and for many more hours of the most intense and rewarding discussions about science and about life, my deepest thanks to all three.

Special thanks to Joe Tietge of the United States Environmental Protection Agency, who helped me to understand the issues involved in this story and who provided continuing perspective and assistance throughout the course of my research.

For the better part of the nearly three years that I worked on this book I had regular, often daily, conversations with Dr. David Gardiner of the University of California Irvine and Dr. Mike Lannoo of the Muncie Medical Center and the Declining Amphibian Populations Task Force. I owe Dave and Mike more than I can possibly say. My deepest thanks for their help, their keen observations, and their unfailing good humor.

Special thanks to Dennis and Rhonda Bock, and also to Eric, Jennifer,

Troy, and Brandon Bock, for their hospitality and their patience with my questions on many occasions.

Finally, thank you to Dr. David Hoppe, of the University of Minnesota–Morris, who opened many doors for me, and whose care for the frogs and the environment in an out-of-the-way corner of the world provided the moral center for this far-flung investigation. Dave is a great teacher and I was privileged to have learned from him.

Introduction

CONSIDER THE FROG.

At the front end of a sleek, sturdy body is a large mouth. At the rear are two powerful, oversized legs. The frog is designed to eat and to escape being eaten. It is a marvel of elemental form and function, one of nature's most successful animals. For 350 million years—since before the time of the dinosaurs—the air on countless summer evenings has been filled with the songs of frogs calling to one another.

Across the chasm of time, the frog has endured. Frogs have survived mass extinctions, ice ages, the shifting of the continents—everything Mother Nature has put in their way. But the frog appears unsuited to the modern age. In recent years frogs have begun to vanish at a disturbing rate, for reasons no one as yet fully understands. Their habitat, obviously, is shrinking. But there is more to it than that, more than just the fact that we are crowding the frog. For even in remote and seemingly pristine places frogs are disappearing. Perhaps this world is dirtier than we have imagined, corrupted by human activity in places that still look so idyllic and unsullied to us that we cannot believe anything is wrong. Or perhaps it is not a consequence of human existence at all but merely nature itself that has decreed the beginning of the end for the frog and set in motion a subtle biological process of extermination, which is invisible to us in every respect but its result—a growing stillness in the night.

This much is clear: The frogs are telling us something. In the summer of 1995, a group of schoolchildren in Minnesota found some frogs whose limbs were malformed. Scientists who investigated the discovery could describe what they saw, but no one could quite comprehend it. The frog, descended in a direct line from the first animals to have four limbs, is the living prototype of all vertebrate architecture. In its design, the frog is a link to our most distant biological origins, and whatever alters that design might, by logical extension, alter us.

Deformed frogs have now turned up in large numbers across the United States and in Canada. In Minnesota, at the epicenter of the outbreak, deformed frogs were found at three separate locations in 1995. By early 1997, abnormal frogs had been reported at nearly two hundred sites in Minnesota, from one end of the state to the other.

From the beginning, scientists viewed this with the gravest concern. Many potential explanations were offered, including the possibility that this was something that naturally happens to frogs once in a great while, something perfectly normal that just hadn't been noticed before. Frogs, after all, have been around more than 80 times longer than humans—we've seen only a tiny fraction of their ecological story. But as more deformed frogs were discovered in more places, the idea that this could be routine occurrence appeared unikely and researchers concentrated on what might be present in the environment that could cause such deformities. One question seemed almost to answer itself.

Why frogs?

In a word, frogs are permeable. Literally. A frog's skin is less a barrier against the outside world than it is an open portal. Frogs can breathe through their skin. Most of the water they need to survive is absorbed through their skin. Whatever is in a frog's surroundings can thus end up inside a frog. All of the terms on which the frog meets its environment are similarly intimate. Frogs breed and develop in water. Their egg masses—translucent, gelatinous globules—are laid in or near ponds and small lakes. Most frogs pass through a larval stage, hatching in the spring and then spending much of the first summer of life—in some cases the second as well—as tadpoles, swimming freely and breathing through gills. At metamorphosis, the larvae sprout limbs, develop lungs, lose their tails, and climb out onto the ground. But they do not go far, dividing the rest of their adult lives between water and land.

Many scientists regard frogs as a sentinel species—a kind of biological early warning system of trouble in the environment. So the one question everyone wanted most to answer is not what is causing deformities in frogs, but rather what it all might mean in the larger scheme of things.

The frogs are telling us something. Is it that the end of millions of years of evolution is reached in the blink of an eye? A deformed frog cannot survive, that much is immediately clear. So deformed frogs, whatever the cause, fall into an already lengthening list of amphibian problems. What will become of the frogs? What is becoming of earth itself?

Biology is not a precise science. It offers ambiguity in place of certainty and complexity rather than simplicity. Trillions of cells make up the human body. Each one is a wonder of molecular engineering millions of years in the making. We are the product of evolution, as is all life that surrounds us. Yet we live in a world now subject to factors so new—artificial chemicals, increasing ultraviolet radiation, global warming—that we have only begun to detect their impact on living things.

The study of life is like life itself: a process of exploration. When I began working on this story three years ago I worried that the ending might come before I was done—that the mystery would be solved and the problem fixed faster than I could record the events. It has turned out quite the opposite. This is a story that will not end, not ever. It will only change shape. We want answers to the mysteries of how life works on this Earth. That's human. But mysteries are also the very heart of life, its one constant feature. There are always some answers. But each comes to us with a thousand new questions. That's biology.

History, the long ebb and flow of evolution, teaches us that life is adaptable, malleable to the vagaries of changing conditions. Earth has seen life explode upon it and all but die out. Life-forms come and they go. Time is like an immense wave breaking on the shore of the present, the narrow place we claim as our own, and we are like grains of sand stirred by forces greater than we can imagine. The nature of life and the character of the organisms that inhabit the planet are not constant. Nothing alive has a permanent claim to existence. We humans are new here. We've only started our run. We want to be careful not to end it prematurely. Suicide is rare, except among our species. We want to pay attention to what we do.

So consider the frog. The frog has sensed a change, a displacement in the order of life as we know it. We may be responsible, at least in part, for causing this change. And we may be the next to feel it. This story is larger than we can imagine. The facts humble us. The evening silence is spreading.

The frog is telling us something. Will we listen?

William Souder

September 1999

And the frogs shall come up both on thee, and upon thy people, and upon all thy servants.

—EXODUS

The clothing of disrepute is diaphanous before any good naturalist's experience.

—STEPHEN JAY GOULD

ONE

The Hell

Henderson

THE MINNESOTA RIVER TRACES A DEEP V ACROSS SOUTHERN MIN-
nesota. From its headwaters at Big Stone Lake on the South Dakota
border, the river runs southeasterly beneath enormous skies, crossing an
open landscape of empty, windswept horizons. In places here you can see
the curvature of the earth. Flowing through scrubby remnants of the great
prairie and down into fertile corn and soybean country, the river follows
the ancient streambed of one of the mightiest watercourses that ever
existed—the river Warren, which 12,000 years ago drained glacial lake
Agassiz. One hundred and fifty miles downstream the Minnesota angles
sharply to the northeast and heads for the Twin Cities, where it joins the
Mississippi. A muddy slurry of silt, farm runoffs, arsenic, lead, selenium,
cadmium, and PCBs, the Minnesota is one of the most polluted rivers in
the country.

But along the second half of its length the Minnesota travels through
a valley that is on either side broad and verdant and flanked by lovely
wooded bluffs. In the summertime the valley is arrestingly lush. A section
above the town of Le Sueur is still claimed by the Pillsbury company as the
"Valley of the Jolly Green Giant." It was near here, on August 8, 1995,

that a group of eight middle school students on a field trip made a shocking discovery.

The day was humid and overcast. It looked like rain. Most Minnesota schoolchildren were still enjoying the last weeks of what had been an unusually hot summer vacation. But at the progressive New Country School in Le Sueur, where classes are in session year-round, a forty-year-old teacher named Cindy Reinitz loaded her nature studies group into a van at 8:30 in the morning. Reinitz and the kids, who were from grades six, seven, and eight, headed north on Highway 93 as a warm drizzle began to fall.

They were going to a farm owned by Donald Ney. Six hundred acres of rich Minnesota loam situated atop the bluff on the east side of the river, the farm overlooks the little town of Henderson far below on the opposite bank. Carved from the woods in the 1850s, the Ney farm is one of the oldest in the area. Don Ney, who was born in Chicago, came here in 1952 to live with his bachelor uncle and two spinster aunts—an arrangement he maintained had put him "under their thumb" for most of the rest of his life. A short, round man in his mid-sixties, Ney was shy but possessed of a ruddy, reticent charm. Never married, his only real passion in life was the land that had finally passed into his hands. More or less. Just before the last aunt died in 1991, she donated 350 acres of the property to Le Sueur County for use as an environmental learning center. Ney figured the gift was probably part of some complicated larger plan to prevent him from gaining total control over the original farmstead. Even so, the result pleased Ney enormously. He was delighted that people, especially children, routinely visited his farm to hike the fields and scramble along its wooded ravines.

Reinitz knew the Ney farm well. She and her family owned forty acres just a half mile to the east, separated from the Ney place by a winding stretch of woodlands and by seven acres of replanted prairie tallgrass that Reinitz frequently wandered over. A tall, attractive woman with long red hair and a purposeful stride, Reinitz planned this morning to take her students on a walk into the woods rising above a creekbed between her property and Ney's. Some of the trees in the area are vestiges of the Big Woods, the towering hardwood forest that formerly covered much of south-central Minnesota. During the half-hour drive from the school, Reinitz watched the leaden skies roiling overhead and wondered to herself what they would do if the rain kept up. The students were a little wild already, arguing over which radio station to listen to and indicating little interest in the passing scenery they were about to explore. It was only Reinitz's second day with this class, and she sensed the kids were testing her.

To her relief, the rain had stopped by the time Reinitz arrived in Henderson around nine o'clock. She drove east over the low concrete bridge that crosses the river. The road curved sharply to the left and the van climbed a steep hill, then turned onto a long gravel driveway that ran along an expanse of undulating fields carpeted knee-high with soybeans. Reinitz drove in about a quarter of a mile and stopped at a field road. She got the kids out and they began to walk. Reinitz surveyed the countryside forlornly, thinking to herself that this trip, which had so often thrilled the third-graders she sometimes taught, might be a complete bore to these older children.

The field road, scarcely more than two well-worn tire tracks, was flat for a short distance. Then it fell steadily as the land gradually sloped away. Up ahead the road bent left, straightened again, and passed beside a massive, solitary ash. Just beyond the tree, in a large, bean-shaped depression, was a wetland. Don Ney could remember when they used to plant that section. But it was always wet and soft and a treacherous place for a tractor. In 1992, as part of the plan to create a nature center, Ney excavated the bottom of this low area, digging out nearly ten acres to create a pond. It is L-shaped, with two arms of roughly equal size and two small, round islands scraped up well out from shore. The pond collects rainwater that washes down the hillsides all around, and several underground drain-tile lines feed in subsoil moisture as well. The banks are a tangle of grasses and cattails and thistle that grows waist-high by late summer.

Reinitz and her class had gone only a few yards from the van when they began to notice small frogs, seemingly hundreds of them, leaping out from under their feet and into the wet, taller grass alongside the field road. The frogs were both green and brown, with darker spots, and looked to be only about an inch long through the body. But their jumps were prodigious, never coming just one leap at a time, but invariably in a blurry sequence of low-trajectory hops as the little frogs surged out of sight. As first one student and then another stuck a hand down to try to catch one, the frogs easily escaped. The kids scampered after them, and Reinitz felt her plan for the day slipping away. It was clear the students would rather do anything than go on the hike she had in mind.

As the class inched forward waves of small frogs continued to part before them as the animals darted right and left, often hard to see except as a sudden, vanishing movement in the grass that lined the edges of the field road. They were northern leopard frogs. Scientific name: *Rana pipiens.* It's a common frog in this part of the country, an important part of the often unseen ecology of a landscape dominated in the hot months by an orderly geometry of crops. Frogs are just one other living part of the yearly cycle in which the land is transformed after a long winter of dormancy. These frogs were juveniles, born that same spring and now newly metamorphosed and dispersing from the pond, spreading out overland in search of insects to eat. They were where they were supposed to be, doing what they were supposed to be doing, repeating an annual pattern that began hundreds of millions of years before the first humans tottered across the face of the earth. Jeff Fish, an enthusiastic, freckle-faced thirteen-year-old who had been literally diving onto the ground after the frogs, caught the first one that didn't look right.

The frog was missing a hind leg. Fish turned the frog over in his hand, thinking at first that the leg must have been bitten off by a predator of some kind. But there was no obvious sign of such trauma—just a continuation of smooth, normal-looking skin over the place where the leg should have been. The boy took the frog over to Reinitz.

Reinitz looked at the frog and at Jeff Fish with dismay. The animal felt cool and a little sticky in her hand as it struggled and kicked its one good back leg trying to escape. She was certain the energetic teenager must have somehow injured the frog in capturing it. But as she knelt and began to examine it more closely a girl in the class brought over another frog— this one with a hind leg that was strangely withered and stiff. Then another student came up with yet another frog with a missing hind leg. Perplexed, Reinitz stood on the hillside, sullen clouds riding low above her, and looked toward the pond. She recalled a visit she'd made to the farm on an excruciatingly hot day back in June, when there had been a great commotion in the pond. She remembered thinking at the time that there must have been large numbers of tadpoles thrashing the water. Now, looking at frogs up on the field that had missing legs or legs that could bend only a little or not at all, Reinitz was amazed that they could get around as well as they seemed to.

"Let's move down closer to the pond," Reinitz told her students.

Sure enough, although there were frogs all over the place there seemed to be more abnormal ones down by the edge of the pond, unable or unwilling to venture away from the water. The kids soon discovered that even frogs with significant leg abnormalities could be difficult to grab. They were quick, and easily navigated the prickly underbrush and the higher reeds at the border of the pond. But after about two and a half hours of chaotic lunging the students had managed to catch twenty-two frogs with their bare hands. Eleven of them had missing or abnormal hind legs.

By the time they had to leave, the students were scared. They talked nervously about the deformities and what they should do. As they walked back to the van, one of the kids asked Reinitz about the cancer rate in the area. Thinking of her own home just through the woods a short distance away, Reinitz wondered about that, too. Mentally she went around the township, thinking of everyone she knew in the vicinity of the Ney farm, and found that she could think of several people with the disease.

Reinitz had another worry. As she talked it over with the students, everyone agreed that they had to report what they'd found right away. But Reinitz didn't think anyone would believe their story unless they kept at least some of the frogs as proof. One of the students hit on the idea of photographing them with the school's new digital camera and posting the pictures on the Internet to see if anyone knew what might have caused the deformities. Reinitz rummaged through the van and found an empty five-quart bucket for Blue Bunny vanilla ice cream. The kids put three of the frogs—two abnormal ones and one that appeared to be all right—in the container and took them back to school. Still shaken, Reinitz began making calls the next day, trying to determine where a person reports having found deformed frogs. Eventually she got ahold of the Minnesota Pollution Control Agency in St. Paul.

I live about seventy-five miles from Don Ney's farm, in country just outside of the Twin Cities that doesn't look much different from the country around Henderson. Our land is bordered on one side by a farm. Each year a crop of corn or soybeans or sometimes pumpkins rises just feet from where our backyard stops. On the other side of the property is a large pond. It's home to ducks and waterbirds and muskrats, plus an assortment of frogs and toads. In recent years it has seemed that the frogs—especially the handsome, strong-limbed *Rana pipiens* I always used to see along the banks—have dwindled. Anyway, I rarely see them anymore.

Like many in Minnesota, my family has an abiding relationship with the outdoors. Despite the ugly outward sprawl of the cities—as lamentable here as anywhere—Minnesota is still mostly open space. The state was widely logged off or converted to farmland a century ago, but plenty of it remains wild today. Our children could name a dozen birds before they could read. They're not surprised to see a turtle excavating a hole in the yard to lay its eggs, or to find a garter snake basking by the swing set. They know the true darkness of the night sky, the frozen fire of moonlight, and what it's like to stand in a forest so quiet you can hear your own pulse. They know when the owls nest and how to look for the white head and tail on a bald eagle as it wheels overhead beneath the sun. We eat the raspberries that grow wild at the edge of the woods and fill our freezer with the fish we catch in the summer and the game we shoot in the fall. We're at ease with nature, and take some comfort in the belief we should be.

This sense of environmental well-being is reasonable. The evidence all around us is that the same force that has done so much to compromise the natural environment—human progress—is also what will save it in the end. A culture that can travel into space and chat on the Internet ought to be able to resolve low-tech issues like pollution or the occasional outbreak of birth defects in a wildlife population. We know what's in a rock from the moon; surely we can know what's in the pond by the side of the road and whether it poses a threat to anything or anyone. For a few decades now, at least, we've begun to police ourselves, restoring the air and water and the earth itself to an approximation of the way it would be in the absence of us.

Or so it would seem.

In fact, the environment is bigger, more complex, more difficult to manage than we might imagine, and the array of government agencies and regulations directed toward that end are far from a perfect answer. In many ways, we do know less about nature and biology than people generally believe. We're also much more tentative about what to do when a new sort of problem turns up.

Cindy Reinitz's call to the Minnesota Pollution Control Agency set in motion a chain of events that would eventually involve scientists and regulators from all over the country. But there was no special reason why this massive investigation started out the way it did.

The MPCA at that time had no particular interest in amphibians; no one on its staff worked with frogs. But someone there had handled a similar call two years before. Reinitz was told to speak with Dr. Judy Helgen, a research biologist in the agency's water-quality division. A provisional employee of the MPCA whose work was funded by outside sources, Helgen's area of expertise was invertebrate biology. Bugs, snails, that sort of thing. At the time she was engaged in a long-term project in which she was trying to develop a wetland index based on various invertebrate species that can serve as "bioindicators" of environmental conditions.

In the fall of 1993, Helgen had looked into a report of deformed frogs near Granite Falls, another town on the Minnesota River, to the west and considerably upstream from Henderson. A resident there claimed to have seen a number of young leopard frogs during the summer that suffered from various limb abnormalities—both extra and missing legs—plus a few that were missing eyes. The MPCA and the Minnesota Department of Natural Resources went out to investigate and managed to catch a handful of abnormal animals. It was sufficient in the researchers' minds to confirm the citizen's report, but nothing more came of it. Analysis of soil and tissue samples turned up mildly elevated levels of arsenic, but that surprised no one familiar with the Minnesota River watershed. Besides, the levels were thought to be below the threshold that would impact the animals. The following spring the frogs in that area appeared normal.

Reinitz was told that Helgen would be interested in what her class had found on the Ney farm. But Helgen was away from the office that week, working in the field with the U.S. Environmental Protection Agency. Reinitz left her a message in voice mail. On August 14, the day she returned to the office, Helgen phoned Reinitz. Each woman would later recall the conversation a little differently.

Helgen says she was immediately alarmed because of the similarity between what Reinitz described and what had been reported at Granite Falls. But if she was worried, she didn't let on. Reinitz thought Helgen sounded more skeptical than alarmed. She says Helgen was polite but didn't seem too impressed because of the relatively small number of frogs the students had caught. Still, Helgen said she would send someone out to have a look, and she did.

The next day, August 15, Helgen asked one of the MPCA's summer interns, Joel Chirhart, to drive down to the Ney farm and see if he could find some frogs. Chirhart, a twenty-three-year-old student worker originally from St. Cloud, had only recently moved back to Minnesota from Texas, where he'd been studying biology. Chirhart couldn't think of any reason why Helgen would send him out to examine frogs, other than the fact that he happened to be available and she was busy.

Chirhart left the MPCA's headquarters in St. Paul around three that afternoon, driving one of the agency's blue Plymouth Voyagers with the words *State of Minnesota* on the side. The weather had turned hot again, and during the ninety minutes or so it took him to make his way south out of the cities and down to Henderson, Chirhart became convinced he was wasting his time. He wondered what was going on. None of it even made sense. Just a bunch of kids who claimed they'd caught some frogs that weren't normal. Looking out over the wilting countryside, Chirhart doubted he'd be able to find any frogs at all on such an oppressive, sticky afternoon.

At the farm, Chirhart went down the long driveway and then turned onto the field road. The blue van rocked slowly down toward the pond under the blistering sun, soybeans fluttering wanly in its wake. Cindy Reinitz was waiting just beyond the big ash tree. Chirhart parked the van, got out, and introduced himself. He and Reinitz started looking for frogs—which turned out to be everywhere in spite of the heat. They caught frogs along the edge of the field road, in the higher grasses, and near the banks of the pond. They caught so many so easily—more than a hundred—that Chirhart didn't even bother to count them. He could scarcely believe what he was seeing. After about two hours, Chirhart walked back to the van and got on a cell phone to Judy Helgen.

"I think you're going to want to get down here right away," he told her.

From Here to There

AN EPIDEMIC DOESN'T NECESSARILY ANNOUNCE ITSELF AT THE beginning. Yet could this really be anything else? For the second time in as many years, deformed frogs had been found in Minnesota. The chances that these were the only populations affected—and that both had somehow been stumbled upon and reported—were remote, to say the least.

Nobody knew how the deformed frogs got there or how long they'd stay or whether more were waiting to be discovered in other places. But the implications of a random encounter with an ecological problem that could be widespread were unsettling. There are frogs—the frogs right there, in front of your eyes that you can see and catch—and then there are frogs in general. They're all over the place. There are thousands of wetlands and fields in Minnesota where leopard frogs live. Within those habitats there are millions of animals. Most of them live and die in reedy inaccessibility, without ever being seen by a human being. So who could really know what was happening to frogs in general? They're out of sight. The discovery at the Ney pond was like the parting of a curtain in a darkened room. Blinding. We're surrounded by nature, but we see only a small fraction of it. We don't know everything. Sometimes it can seem that we know next to nothing.

On August 18, the Friday after Joel Chirhart's visit to the Ney farm, Judy Helgen went back down with him to see it for herself and to try to collect another sample. She went in her official capacity, representing the Minnesota Pollution Control Agency, which ostensibly represents the interests and safety of the Minnesota public at large.

Chirhart's phone call from the banks of the Ney pond had disturbed Helgen more than her first conversation with Cindy Reinitz. "At first I didn't quite get what he was trying to tell me," Helgen told me later. She didn't think it was possible. "I thought he had the numbers mixed up." The situation, if Chirhart had it right, sounded worse than what had been seen at Granite Falls. This time there were a great many abnormal frogs and no obvious source of chemical contamination near the pond. The Ney farm is within a mile of the Minnesota River but nearly two hundred feet above it—close enough, but with no chance of the waters of river and pond mixing.

About one of every four frogs Chirhart had caught was afflicted with some kind of leg abnormality—and not just the missing or nonfunctional legs Reinitz and her students had found. The range of malformations Chirhart reported was amazing. All the abnormalities arose in the hind limbs. There were frogs with extra legs, singly and in pairs. Some of the additional legs were fairly normal-appearing duplicates growing alongside the primary limbs, others were translucent pink and undersized, sprouting at odd angles from the animals' hindquarters. There were fleshy masses and tapering, undifferentiated protrusions that could not quite be called legs. Some frogs had legs that appeared shrunken and dessicated and stiffened—like jerky, as if they had been cooked to the bone. Others had legs that seemed to split in two somewhere along their length, or legs that were weirdly bent and inflexible, with feet pointing every way but the right way. Some legs were twisted like corkscrews. There were legs that looked normal but had more than one foot growing from the end. A number of the frogs had paralyzing skin attachments—like the webbing between a duck's toes—that spanned between the upper and lower leg segments, preventing the frog from fully extending its hindlegs, making swimming and leaping difficult or impossible. Frogs that were missing legs were like optical illusions—so natural and unperturbed was the skin along the flank where the leg should have been that it almost seemed as if the it was meant to look that way. "It was like science fiction," Chirhart told me later. Helgen's initial skepticism vanished when Chirhart arrived back at the MPCA with a large sample of the deformed animals. When the state van again bounced down the field road, passed underneath the big ash tree, and came to a stop near the pond, she became even more worried.

Helgen's first impression of the Ney pond was how orderly it looked. The regularity of its banks, the perfectly round, domed islands scooped up from below, and the scant emergent vegetation all gave it a decidedly artificial appearance. Helgen pulled on her waders and walked into the water. Near the shore the soft bottom sucked at her boots, but presently the mud gave way to a harder, shallow, surprisingly level bottom farther out. It felt like smooth clay. "It's hard to put into words," Helgen told me later. "It just looked and felt so manmade. It felt quite unnatural."

Helgen made a mental note of her reaction. A shallow wetland like this ought to be densely fringed with aquatic vegetation. And the water should be transparent. Healthy wetlands are great natural water filters. Until you start walking around in them and roiling things up the water is almost always clear. But Helgen thought the Ney pond looked like muddy tea. It was choked with clouds of unrooted, submerged weeds. When she talked with Don Ney about how the pond had been excavated her suspicions grew. Bulldozing would account for the hard-packed bottom. But it might also mean that something toxic that had been long buried in the sediments had been stirred up into the water.

Onshore, frogs were everywhere. And just as before, many of them had abnormal legs. Helgen noticed that some of the deformed frogs appeared to be doing all right. They were about the same size as the normal ones and they were very active. But they moved awkwardly, and Helgen winced at the frogs' tortured attempts to escape capture. Frogs that were missing legs, or which had only parts of legs, would tip or sometimes spin as they jumped, often flopping over, their bellies coming up white in the sun. Animals with more profound abnormalities crawled along on their front legs, some barely able to move. Helgen saw raccoon tracks around the pond and wondered to herself how many of the frogs she was looking at would starve to death and how many would be eaten. "It was clear a lot of them would be picked off by predators," she told me.

Helgen said something like a third of the frogs they caught that day, maybe more, had leg abnormalities. Helgen told Reinitz and Ney she would be back, and over the next several weeks she visited the pond on numerous occasions. On September 21, in the final days before the frogs would begin moving toward their overwintering sites, she arrived with

Robert McKinnell, a biologist from the University of Minnesota who had studied frogs, mainly northern leopard frogs, for many years. Helgen had previously consulted with McKinnell about the Granite Falls report. He'd been skeptical that time and never actually visited the site. This time would be different.

"I'll come," McKinnell had told Helgen when she called him about the find at Henderson. A distinguished senior member of the university's faculty, McKinnell was better known as an expert on cell biology and cancer. But those studies had put him onto frogs back in the 1950s. Frogs were easy to come by and they got cancer. Plus they lived in places that McKinnell, an otherwise urbane man, liked to visit.

Frogs also intrigued McKinnell on their own. He was particularly interested in the origins of two local color variants of the leopard frog. Few people in Minnesota knew more about the species than McKinnell, and despite his advancing years and buttoned-up appearance, he was a spry and able field biologist who quickly sized up the situation at the Ney pond. He didn't like it at all.

Standing by the shore of the pond, McKinnell struggled to make sense of what he was seeing. It was, he recalled later, a pretty day, very summery. Frogs were not immediately evident, but sure enough, as he scuffed around in the high grass, McKinnell saw them springing out from under his feet, angling for cover. Frogs can see a net quite well and will dodge one with suprising ease, especially if it's brought down from above. Often the best way to catch a frog is to get the net down in front of it as your foot comes up from behind, so that it leaps in on its own. Maneuvering his net this way and that to cut off the fleeing amphibians, McKinnell worked his way through the tangle of weeds and thistle, pausing now and then to extract a frog caught pulsing and struggling in the mesh.

McKinnell started out on high ground, away from the water. At first, most of the frogs he looked at were normal. But after catching about thirty he finally came up with one that was missing a leg. He moved down closer to the pond and right away began to catch more abnormal frogs. It was obvious that the unaffected animals were moving out, dispersing normally across the surrounding fields, while the deformed ones were stuck in or near the water.

McKinnell later recalled that about 20 percent of the animals he found that day were abnormal. Only a few had extra limbs. Some had legs that were brutally twisted and misshapen. A number were crippled by skin webbings conjoining their leg segments. McKinnell was especially struck by how many missing and partial legs he saw, and was at a loss as to what might be the cause. Just as Cindy Reinitz and her students had done a few weeks earlier, McKinnell considered the possibility of injury. His own experience was that although once in a great while you might encounter a frog that has lost a leg to a predator or in some sort of accident, it was rare. In years of collecting and examining frogs he had hardly ever seen one missing a leg.

On the other hand, McKinnell had on countless occasions witnessed the act of predation. Like most biologists who've spent a long time in the field, McKinnell knew the possibility of escape all but vanishes once a predator has gotten ahold of its prey. The fact of life in the wild is that predators do not, as a rule, perform amputations. Predators are not nibblers. Once an animal's leg or any other body part goes into a predator's mouth the rest of the animal is all but sure to follow. McKinnell had often seen garter snakes eating frogs. "You think the frog is going to get away," he told me later. "But it doesn't. Bit by bit, the snake swallows the frog. It goes in. It goes down. Not part of the frog. The whole frog."

The idea that failed predation could account for missing limbs among a large percentage of an entire population of animals struck McKinnell as presposterous right from the start. It seemed far more probable that the frogs at the Ney farm were survivors of something else altogether, some catastrophic developmental miscue, or possibly even a genetic mutation. Just as they had in Granite Falls two years before, the researchers immediately suspected chemical contamination. McKinnell, half-joking, wondered aloud whether someone might have dumped a can of paint thinner into the pond.

Of course, it was also possible that limb abnormalities in frogs are just something that happens from time to time and that the students had merely happened upon a rare natural occurrence that would subside on its own. In nature, McKinnell often noted, strange things sometimes happen just because that is the way nature works.

But what if such a strangeness were to spread?

The find at the Ney pond did not disappear as the outbreak at Granite Falls had. Instead, all through late August and into September, more and more deformed frogs were found there. Then they began showing up in other parts of the area near Henderson. Some turned up in a pond on a farm just north of Don Ney's. Then more were discovered in a small, oval-shaped pond on the southern end of Henderson itself, down in the valley and on the other side of the river in Sibley County. A family stopped at a wayside rest area on the Rush River, a tributary of the Minnesota just west of Henderson, and found what they later claimed were many "grossly deformed" frogs. In early September, Reinitz and her students caught the first animal with multiple front legs—four on one side—at the Ney pond. Not long after that Cindy Reinitz found a leopard frog on her own property that had one eye and three hind legs.

Reinitz and her students began developing a Web site where they posted news and photos of the deformed frogs of Henderson to Internet users. Meanwhile, Helgen and McKinnell looked for some common denominator among the various sites that might provide a clue to the cause of the deformities. As more deformed frogs were found, the possiblity of genetic mutation seemed increasinngly unlikely. Nobody thought it plausible that such a random variation would appear simultaneously among multiple populations at multiple sites. Meanwhile, Helgen and McKinnell thought all of the places with deformities were similar to one another in some telling ways. Most were farm ponds—wetlands that had been disturbed by excavation or livestock watering. And all of these seemed at least marginally associated with the Minnesota River.

Then, on September 29, Judy Helgen's phone rang again.

This time the caller was a teacher at an elementary school in Litchfield, a farming community an hour's drive due west of Minneapolis and about forty miles northwest of Henderson. The teacher said one of her students had brought in a frog that had no hind legs at all. The student said he'd gotten the animal from a neighbor's pond where he often went to collect frogs to use as bait when he went fishing.

The pond belonged to Mike and Audre Kramer, who live on a couple dozen acres of gently sloping land on the North Fork of the Crow River. They have a huge lawn and lovely, sprawling flower gardens, all surrounded by crop fields that angle down gradually to the river. There's a farm across the road and slightly uphill. The pond is manmade and about forty yards from the house. It's seventy-five feet long, maybe fifty feet wide at its broadest point, and shaped roughly like a pear. Its banks are steep and firm. The Kramers dug it out in 1993, but a spring feeds into the pond and an overflow runs out one end and through a long, shallow culvert that leads downhill toward the river. Audre stocks the pond with turtles, painting their backs with her grandchildren's initials. Audre told me that when the first carload of scientists got there they looked disappointed. "It's really not a very impressive pond," she said.

But Audre was impressed with what she was finding in it. She'd been saving some of the legless frogs in an aquarium. They were the first thing she wanted to show to her visitors—one of whom, Bob McKinnell, politely went over for a look. Kramer could sense his mood change as he looked at the frogs. McKinnell peered down into the aquarium for a long time. Then he straightened and motioned to Kramer to walk with him a little way. "I think," he told her quietly, "that you should prepare yourself for some attention from the press."

Just like the Ney wetland, this was a recently excavated pond. McKinnell and Helgen thought this was immensely relevant—the one seeming consistency among all the sites they were now looking at. Since the Kramers were nowhere near the Minnesota River it seemed pretty well eliminated as a factor. Also, just as at Ney, these were young-of-the-year leopard frogs. Many of the frogs in Audre's aquarium were missing both hind legs. Down at the pond it was a horror show: The banks were alive with gruesome clots of crippled leopard frogs. The scientists collected ninety-four animals and ninety-one of them had no back legs. Most of them were stranded at the water's edge, struggling and dragging themselves around in the mud.

News of the discovery at the Ney farm spread quickly. A television station from the Twin Cities visited the pond and broadcast scenes of Reinitz and her students collecting frogs. Accounts of the deformities made a few headlines in newspapers around Minnesota. One longer story that appeared in the Minneapolis *Star Tribune* in early September quoted Judy Helgen, who by then had launched an investigation. Helgen said high rates of deformities in a population of amphibians appeared to be without precedent and that the situation on the Ney farm was "beyond extraordinary." The story also noted that amphibians are considered "indicators" of environmental degradation because they are "highly vulnerable to toxins." Helgen speculated that a chemical contaminant might indeed be behind the malformations, but cautioned that it would require some time to determine what it was. The inquiry, she said, was likely to take "at least several months."

Almost all of this was either wrong or wildly optimistic, as everyone involved soon enough discovered. Outbreaks of frog deformities were not unknown to science. Nor would the current mystery be solved in anything like a matter of months.

Helgen's hopes for an easy answer faded rapidly as more deformed frogs started turning up in other parts of Minnesota. Each new find sparked interest—McKinnell was right about the press, which eagerly covered the outbreak, although none of it seemed to cause any serious alarm. The tone of the coverage was sober, but the public response was only general bemusement. The frogs were widely referred to in classic sci-fi terms as "mutants," even though it was by then understood that this was technically not the case. People joked that extra-legged frogs were a boon that would make Minnesota a leading supplier of frogs' legs to the restaurant trade.

I followed the story with some interest whenever it resurfaced, but I found the articles were curiously devoid of elaboration. Most of the reports dutifully noted that the MPCA was performing "tissue analyses" on the affected frogs. For what? Chemicals, especially pesticides, were usually mentioned. This was understandable enough. About 50 percent of Minnesota—some 26 million acres—is agricultural land. Farmers annually apply upward of 14.5 million pounds of herbicides and insecticides to the corn crop alone in the state.

But it scarcely seemed possible that anything approved for use on human food sources would cause such grotesque wildlife abnormalities. It was the end of a Minnesota summer and my family was eating the season's produce, especially sweet corn, in volumes. Surely it was safe. Looking out at our own pond I thought about the frogs and also about a line from *One Fish Two Fish, Red Fish Blue Fish*, a Dr. Seuss book that I always enjoyed reading to my kids. It went, "From here to there, from there to here, funny things are everywhere."

What was actually happening, however, was no joke. In late summer of 1996 I went to see Bob McKinnell. There's almost always more to a story as time goes by; this time there hadn't been. Deformities continued to be reported sporadically in the papers, but a year after Cindy Reinitz's anguished call to the MPCA the "investigation" remained largely undescribed to the public. And apparently it wasn't getting anywhere.

This was puzzling. Why was it taking them so long to figure out what had caused the deformities? Why couldn't they simply analyze the water to see what was in it that shouldn't be there? And who were "they" anyway?

McKinnell greeted me at the elevator when I stepped out on the seventh floor of the Biological Sciences Building at the university's St. Paul campus. He led me down a long, fluorescent-lit hallway to his office, a narrow institutional-looking warren piled floor-to-ceiling with books and scientific periodicals. A tall window looked out over the Minnesota state fairgrounds, which are adjacent to the campus.

Silver-haired and formal, McKinnell was nearly seventy and approaching retirement. He struck me as a bit of a dandy, although there was a flinty sharpness in the way he spoke. He was originally from Missouri, but over the years he'd acquired the clipped consonants and intermittently wooden syntax that is quintessentially Minnesotan. We talked for a long time. His speech was formal, occasionally technical, but laced with odd colloquialisms. He'd begun studying tumors in frogs while in graduate school at Minnesota in the early 1950s. In the 1960s he surveyed frogs across the state—out in the "boony docks," he called it—under a grant from the American Cancer Society. In the course of his travels he'd gotten to know many of the area's longtime commercial frog "pickers." In those days, someone sitting down to a dinner of frogs' legs in a restaurant in Chicago was most likely eating *Rana pipiens* from Minnesota. The processing procedure was simple, if gruesome. The legs were snipped off the live frogs with shears and soaked in water overnight. This made them swell to double their original size before freezing. The bodies were discarded and later sold to mink ranchers as feed. McKinnell would hang around and open the abdomens to inspect them for tumors, which turned up fairly often.

When I asked him if the deformities at the Ney farm were unusual, he said he had doubted it at first. "I thought surely this was no big deal," he said. "You find abnormal frogs wherever you find frogs. There's nothing new about the occasional abnormality in nature. You can go to the state fair and see a pig with two heads."

I looked idly out the window toward the fairgrounds as McKinnell continued on. He explained that Judy Helgen had described the scene at the Ney farm to him in such dramatic terms that he had had to see it for himself. He said he changed his mind quickly once he visited the Ney pond. In forty years of looking at frogs, he'd never seen anything like it. "Let me show you," he said.

We walked across the hall to McKinnell's lab and looked at a number of small leopard frogs that had been collected that season. They were in plastic containers lined up on a table and looked lively enough despite their sundry twisted or missing limbs. McKinnell took the top off one box that had a single frog in it. "What do you see here?" he asked.

I looked at the animal and couldn't see anything wrong with it as it rested on a bed of moss. All four legs looked perfectly normal. Finally, I saw something unusual against the background of dark spots on green skin.

"It's missing an eye," I said.

McKinnell smiled and reached into the box, expertly snatching up the frog as it ricocheted from wall to wall.

"We thought so, too," he said. "Then one day we were feeding it and we discovered this."

McKinnell pulled the frog's mouth open and showed me what was inside. It was the other eye, growing at the tip of a stalk of flesh that extended down from the roof of the mouth.

"Just how widespread do you think this is?" I asked him after a moment.

"Oh, good grief," McKinnell said. It's one of his favorite expressions. "You really don't know, do you? Come look at this."

He took me over to a large map of Minnesota taped to the back of a door. McKinnell and the MPCA had been busy tracking reports from all over, most coming in from private citizens. There are eighty-seven counties in Minnesota. More than fifty of them had been shaded in on the map during the past twelve months. McKinnell gazed at the map, slowly shaking his head.

"That's where we've found them so far," he said. "But that's really just an indication of where people have looked. Personally, I think they're everywhere."

3
Creepy Crawly

 HE TOWN OF BRAINERD LIES ONE HUNDRED MILES TO THE NORTH
of the Ney farm in a "transition" region where the open grasslands and
agricultural fields of the south give way to the northern pine forests, and
Minnesota begins to look like the Minnesota you see in postcards. Brainerd is known to summer tourists and fishing enthusiasts as the capital of
Minnesota's prime lake vacation country. The town nestles alongside the
upper Mississippi River, at the southern edge of a great splash of sylvan
lakes to the north and east, big waters like Gull and Pelican, Leech, Ten-
mile, and the Whitefish chain, plus hundreds of smaller lakes stippled
across the landscape. Fifteen miles to the east of Brainerd is the immense
round pan of Lake Mille Lacs, the state's legendary walleye lake. The area
teems with visitors year-round, from the sultry laze of deep summer into
the frigid envelope of winter, when the skies are opalescent, the pitched
whine of snowmobiles keening through the forests travels for miles on the
frozen air, and anglers plow roads out to small rainbow cities of brightly
painted ice-fishing houses in the middle of Mille Lacs. Water, liquid or
solid, is the focus of everything here. Brainerd was one of the last towns in
America to fluoridate its drinking water, something the citizens resisted for
years in the belief that nature could in no way be improved upon.

Dennis Bock was born in Brainerd in 1951 and grew up there just blocks from the Potlatch Corporation, a paper company where he now works in the coating department. When he was eighteen, Bock's parents moved out to the country east of town, near Nokay Lake and the place where Dennis's grandfather had built a homestead early in the century. About the time they moved, Dennis met Rhonda Wiltz, a quiet, pretty brunette who was in the ninth grade. They didn't date until some years later, but Dennis recalls that he "kept track of her" all along.

In the fall of 1977 Dennis and Rhonda became engaged and started looking for some land near his parents' place. There didn't seem to be much around that they could afford. Out of the blue Dennis's grandmother offered to sell them a small section of her property. The deed claimed about thirty-five acres, but one stretch of the boundary ran through a lake bordering the east side of the property, and back in the 1930s a dairy farm next door had encroached from the north. The actual remaining area of dry land worked out to something like twenty-six acres. Dennis had always secretly hoped they'd end up there some day. He and Rhonda were thrilled.

The Bocks began clearing the property for a house that fall and started building the next summer after their wedding. They moved in finally in 1979 at Thanksgiving and have lived there ever since. Rhonda, who used to work as a nurse, has tended the home and their growing family—Eric, Jennifer, Troy, and Brandon—plus Buster, a plump, good-natured golden retreiver, and a cat with no name that skulks about and jumps into cars whenever anyone drives up and opens a door. Dennis works rotating shifts at Potlatch. He's a wiry, intense man, with a gentle manner and a lively, agile mind. He's also fastidious, and every inch of the property and the house is always in perfect order.

For years, the Bocks lived the kind of dreamy, unruffled middle-class lives that most of us imagine as rare yet somehow archetypally American. A big part of this picture was the timeless symmetry of the rural setting. The Bocks' house—a sunny yellow five-bedroom split-level—faces south at the end of a long, flat, gravel drive. It's surrounded on three sides by a large yard with neatly tended flower gardens front and back, and an ample vegetable garden near the woods.

About sixteen acres of the property are leased to the dairy farmer next door, who plants clover or alfalfa and sometimes corn on either side of the driveway and out to the west edge of the property. In the summer of 1997 the Bocks made about $30,000 worth of improvements, including a blacktopping of the parking area near the house and the addition of a large sheet-metal pole barn where Dennis keeps his boat and the four beloved snowmobiles the Bock family tools around on in winter.

The east side of the property is wooded, a mix of oaks, maples, and pines mainly. A swath of lawn runs from the house down a steep hill to the bowl of the lake. The lake isn't named. But it's pretty good-sized, roughly thirty acres in area, and almost perfectly round. There are no other homes on the lake, no other manmade structures of any kind visible from the water. The woods ring the shoreline seamlessly, except for a spot about a hundred yards north of the Bocks' where the cows come down to the water. There's an electric fence extending out into the water to keep the herd where it belongs. The bank there is open and softly trampled down.

The lake is spring-fed, the waters brownish but relatively clear over a muddy bottom that is easily roiled. Lily pads and reeds grow out from the shore and rock gently in the waves when the wind kicks up. Dennis and Rhonda say they've been all over it and have never found any spot in the lake to be deeper than about ten feet.

The Bocks have a twenty-foot dock and keep a two-person paddle-boat beached alongside it. For most of the time they have lived here, the kids swam and played in the lake, and the whole family for many years enjoyed fishing for sunfish, which they love to eat. Usually the fishing was good, though sometimes the panfish were less plentiful after an especially harsh winter. Dennis and Rhonda have often seen eagles taking larger fish from the lake, too.

The Bocks put their first well near where the vegetable garden is. It was fifty-seven feet deep, drawing water from a subterranean layer of fine sand. The well caused the Bocks nothing but headaches. Like much of the well water in this part of Minnesota, theirs was loaded with iron—so much iron that a heavy, rusty sludge filled the pipes in the house and the tap water ran red. In 1988 they sank a second well, right alongside the first, that went down ninety feet and terminated in a layer of coarse gravel. This was an improvement. Although there was still high iron content in the Bocks' water, their water softener made it clear and, so far as they knew, safe.

One day in July of 1995—about a month before Cindy Reinitz and her nature studies class visited the Ney farm—Dennis and Rhonda were sitting at the dining room table talking with friends and keeping a casual eye on the kids, who were outside. The Bocks' dining room overlooks the lake. It was late afternoon. As the adults chatted, Brandon, who was ten at the time, came up to the window holding something.

"Look at this weird frog!" he said.

The Bocks can't remember whether the frog had five legs, or maybe it was six, but Dennis is pretty certain it was a leopard frog.

"We didn't even really check it out," Rhonda told me when I talked with them two years later. "I guess we just assumed it was some minor freak of nature."

Dennis and Rhonda promptly forgot all about the frog. Even when they heard the news about the outbreak at Henderson a few weeks later, Brandon's discovery seemed little more than an ironic coincidence. But then in late August the kids found more deformed frogs—about a dozen or so. All of them were leopard frogs. All were juveniles. And all had extra hind legs. Dennis decided to take them to the local office of the Department of Natural Resources. He asked the kids to find something to put the frogs in, and they did—an empty ice-cream bucket.

The next day, Dennis took the frogs with him to work and showed them to his buddies at Potlatch. They were not impressed. They told Dennis they saw frogs like that "all the time." Figuring he'd make a fool of himself if he went over to the DNR, Dennis took the bucket out behind the plant, turned the frogs loose, and went back to work.

But then came more reports of more deformed frogs from other places around Minnesota. Rhonda felt they should contact someone, and finally she did call the local DNR office. The woman at the DNR said there was nothing their department could do about the frogs, but that she would pass on the information to someone who could. It was late August.

A couple of days later the Bocks got a call back. Dennis answered. The man on the line introduced himself as David Hoppe, a herpetologist at the University of Minnesota campus at Morris, about two and a half hours away from Brainerd. He explained that he was consulting with the MPCA on the investigation into the deformities and that the agency had asked him to follow up on what they had found.

Hoppe was very polite, but he seemed skeptical and asked Dennis a lot of questions—about the lake, about what the deformities looked like, about what kind of frogs they had been catching. This had become the pattern by now. Like McKinnell and Helgen before him, Hoppe found the descriptions of deformed frogs so improbable, so unlike anything he'd ever run across in years of fieldwork, that he couldn't quite get himself to believe what he was hearing. Even as the reports of deformities began to pile up, the researchers struggled with doubts that this could really be happening. Each new outbreak was greeted the same way: It can't be.

Dennis Bock got the impression that Hoppe thought his story was strange. He thought Hoppe sounded ambivalent about coming up to see the lake for himself. But they talked some more, and after awhile Hoppe said that it did in fact sound as if they were seeing some of the same things that were being reported elsewhere that summer. But he wasn't just going to drop everything and rush up to Brainerd. Hoppe said he was busy with the start of the school year. He would try to come up in early October.

————————————

David Hoppe couldn't remember a time when he wasn't in love with animals. As a kid growing up in Pine Island, a rural crossroads in south-central Minnesota, he was always catching something alive to keep awhile. "Snakes and stuff like that," he told me once in a dismissive way that suggested there wasn't anything else of interest to relate on this count. Hoppe doesn't go on about himself, or about much of anything else.

Hoppe's fascination with living things didn't at first appear to be a calling. He started out studying math and engineering in college, at the University of Minnesota in Duluth in 1960. He wasn't happy. But then he got hooked on biology after taking a class in it his sophomore year. Hoppe transferred down to the St. Paul campus where he could pursue studies in zoology. He planned to specialize in mammals, but diverted into herpetology when he became intrigued by the puzzling "polymorphisms," notably color variations, that occur in some frog species. For his doctoral work at Colorado State, Hoppe investigated the relationship between behavior and the different skin colorations that occur in chorus frogs in high-altitude habitats.

If this sounds like the sort of work you always imagined biologists doing—forget it. Hoppe's résumé might have been fairly typical once. Not anymore. Biologists who study animals—this sounds crazy—are on their way out, and have been for decades. The image most of us have of biologists as natural historians working in the field with binoculars and notebooks is a sepia-toned anachronism. Ever since Watson and Crick first described the double helix of DNA back in 1953, the field of biology has looked ever inward, ever more deeply into the chemistry of life as the source of all understanding. Biologists found that they could study virtually every aspect of life—from development to disease, from phylogeny to evolution to morphology—at the subcellular level. Molecular biology, not "whole organism" biology, is the mainstream now, the dominant precoccupation of the entire field of life science. Zoologists? They're an endangered species. Herpetologists? They're dinosaurs.

But a dinosaur was exactly what the fledgling deformed frog investigation needed. By the time Brandon Bock picked up an abnormal frog in his backyard there were dozens, probably hundreds, of so-called gene jockeys in Minnesota—biologists and graduate students who could splice and clone DNA, including DNA from a frog. But there were only a handful of biologists from the old school, scientists who could actually find a live frog in the middle of a swamp, let alone tell you what species it was or anything about its life cycle. Not many biologists could tell a frog from an insect by its call, or knew when they would breed, or could even distinguish between the male and the female if they held one of each in their own hands. Everybody knew a frog should have four legs; hardly anyone remembered how many toes.

For the record, it's eighteen—five on each hind leg and four on each one in front.

Hoppe later talked to me about this ongoing intellectual attrition, in which whole-organism biologists are simply fading out of the picture, and along with them much of what we used to know of the natural world. Natural history is now to science what fresco painting is to art: a beautiful tradition that hardly anyone practices anymore. "It's obvious in a lot of colleges, especially smaller ones," Hoppe said once. In his first job, at tiny Mayville State in North Dakota, Hoppe had been the entire biology department. "What happens," he said, "is that people move on and they take what they know with them. When the herpetologist retires they hire a cell biologist. When the ornithologist retires, they hire a molecular biologist. Then the mammalogist gets disgusted and goes somewhere else. After that the ecologist does the same. And so on."

"So now I suppose all of your students want to be molecular biologists," I said.

"Oh no," Hoppe answered quickly. "Most of 'em want to go into medicine."

Hoppe seemed to regard these twilight times for the study of natural history as a regrettable trend, but something he'd long since come to terms with. I think he'd much rather not be holding down a place near the end of the line for herpetology, but I doubt that he loses much sleep over the possibility.

I had met Hoppe in the late summer of 1996, a few days after my visit to Bob McKinnell. The two men knew each other well. They'd been collaborators for some time, thanks to their mutual interest in the Burnsi and Kandiyohi morphs—two variants of the leopard frog that live in parts of Minnesota. For years they'd been working on the problem of explaining why these amphibians persisted despite their relatively small numbers and narrow range. The Burnsi morph is a bronze version of the leopard frog that has no spots. The Kandiyohi morph looks closer to the original, but its spots blend together in a reticulated pattern. Both differ from the standard leopard frog—what a field biologist would simply call the "wild type"—by a single gene.

Hoppe's office on the first floor of the Science building was about the size of your average master bedroom. It was connected to a similar-sized space next door that served as his laboratory, which was filled with aquaria and an assortment of "herps." Frogs. Snakes. Salamanders. There was the inevitable mountainous clutter of books and journals jamming bookcases and piled up along the cinder-block walls in the office. Hoppe's Brittany spaniel, Debbie, dozed leadenly in an undersprung armchair near his desk. Above all this an assortment of stuffed-and-mounted aquatic "trash" species hung on the west wall. Eelpout. Bowfin. Sea lamprey. Hoppe said he liked having these around because they were so universally detested by everybody else.

Hoppe was fifty-three at the time. He was of average height, solidly built, and had a shock of brown hair that was graying and thinning that he absentmindedly pushed at with his hand as we spoke. Hoppe was cordial and responsive to my questions without being noticeably talkative. When I got to know him better, Hoppe turned out to have a penetrating mind and a quick, devilish sense of humor. But he's a guarded, intensely private man who opens up slowly. I came to see that he was well suited to where he was—in a small college in a small town in a small, faraway corner of Minnesota. It's always hard to tell whether someone's work is an extension of their private life or whether it's the reverse, but to someone accustomed to long, lonely hours in swamps the scale and the pace of a rural college town is perfect. Hoppe seemed to me an inherently contented individual who'd shaped his life to ensure he stayed that way.

Contentedness sometimes reinforces reticence, and that proved to be the case with Hoppe. A divorced father of two, he now lived away from Morris, on a quiet bay of Lake Mary near the resort town of Alexandria, in a home he shared with his companion, Carlene Ness, whom he'd known since grade school. When he wasn't working he was most often at home fishing or on the road with his daughter, a standout high school athlete, at one of her basketball games. I've known him for four years. He never told me much of anything else about himself.

Hoppe did tell me right off on that first visit to his lab that he wouldn't speculate about what might be causing the deformities. "I'm trying to remain objective," he said. "My job is to describe the biology of the situation. Which species are affected. Where this is occurring. What the frequencies of abnormalities are. What types of deformities we are seeing."

Hoppe said about all he was sure of at that point was that the deformities were not genetic mutations, but rather developmental abnormalities. He also seemed pretty certain that the "biology of the situation" was unusual, to say the least. He indicated that limb abnormalities in amphibians had been reported in the scientific literature going back a long time. But he thought that what was happening in Minnesota was something that hadn't happened here before. "I don't think anyone in Minnesota has looked at more frogs than McKinnell and I have over the past twenty years," he said. "We never saw anything like this."

Hoppe was dubious when he got off the phone with Dennis Bock that first fall of 1995. Bock sounded like a reasonable, intelligent person who must have seen what he claimed to have seen. But it was still so incredible. Hoppe had been following the news from Henderson from a distance, much as he had listened to the report from Granite Falls in 1993. Back then Hoppe had been contacted by the MPCA to help conduct a field investigation. But he never saw anything more than a blurry picture of some dead frogs in a bucket, and when he personally surveyed the site the following year he saw only normal frogs. This was, in Hoppe's view, hardly surprising. In all his time field collecting, Hoppe had never seen anything other than an occasional minor deformity among the thousands and thousands of frogs he'd handled—certainly nothing that affected a large percentage of an entire population.

In fact, Hoppe and McKinnell had gone out of their way a few years before to suggest that the appearance of abnormal animals in nature is generally too rare to worry about. It was in a brief aside in an article on color variations in leopard frogs they published in a conservation magazine, *The Minnesota Volunteer*. "We are not looking for two-headed, six-legged, or otherwise freakish offspring born in ponds contaminated by radioactive fallout or toxic waste," they wrote. The year was 1991.

Four years later, Hoppe found himself contemplating the unthinkable. Although he was determined to remain neutral on the question of causes, Hoppe privately suspected that Helgen and McKinnell had good cause to suspect chemical contamination. He knew that in the days before commercial garbage hauling and sanitary landfills came to rural Minnesota, people disposed of everything in whatever hole or ditch or crease in the woods was available and out of sight. He couldn't imagine a frog pond anywhere in the state that wasn't within a mile of some old dumping area. Contaminated groundwater and pond sediments were probably pretty common. Although this didn't quite rise to the level of a working hypothesis for Hoppe, the idea was there, in the back of his mind, all the time.

Hoppe finally broke away from his teaching schedule one day during the first week of October. The long drive over to Brainerd was discouraging. Summery weather had held through most of September, but it had lately turned cooler. Hoppe hoped the day would warm up enough that he might find frogs congregating near the water, getting ready to hibernate. But there was a rawness on the wind now that made him wonder if he hadn't waited too long to go. When he pulled up in front of the bright yellow house he knew he had. The trees were pitching in a stiff breeze that pulled leaves from their branches in long swirls of autumnal color. Overhead the sky was gray and cold. Hoppe shivered as he stepped out of his car.

In an instant he forgot all about finding frogs. Instead, he stared in disbelief at the Bocks' lake.

Hoppe had expected some approximation of the farm ponds that Helgen and McKinnell had described to him weeks earlier—classic leopard frog habitat. Instead he found himself looking out over a nearly pristine, natural lake, maybe half a mile across and obviously deep enough to harbor a variety of species through the winter.

There wasn't a frog to be seen anywhere that day. Hoppe spent a long time talking with the Bocks, whose story held together, growing more credible and more detailed as he listened. The Bocks liked Hoppe right away. His straightforward, unpresumptuous manner was disarming. Hoppe seemed like a regular guy, with his jeans and his quiet way and the neatly combed hair that he kept trying to push down in the breeze. They were surprised at how eager he now seemed to investigate their lake. He asked for their permission to make it a study site in the future. Dennis and Rhonda told me later that it was almost as if Hoppe knew instinctively how important their discovery had been.

"He seemed to think our lake was unique," said Dennis. "That it was dissimilar to the other places they were exploring."

"Isolated," said Rhonda.

"Dave seemed real excited about that," Dennis added.

The Bock children had caught the deformed frogs on the lawn, and certainly they were leopard frogs. Hoppe explained that their lake should have some other species in it—like mink frogs and green frogs—that rarely venture out of the water. As disappointed as he was at not finding frogs that day, Hoppe told the kids to be on the lookout. If the weather warmed suddenly the animals might yet reemerge and he wanted them to call him immediately if that happened. Hoppe told the Bocks that whether the frogs showed themselves again that fall or not he'd be back in the spring for sure.

Dennis and Rhonda wondered how worried they should be. Although she'd been the one who insisted on reporting the deformities, Rhonda still thought they were most likely just a fluke—one of nature's occasional mistakes. Dennis wasn't so sure. He couldn't stop thinking about the old township dump.

There were at least two possible sources of ground contamination in the area. One was a small area near the south end of the lake where a former neighbor had disposed of trash, including batteries and other automobile parts, over a period of years. Dennis thought that didn't seem like too big a deal. The township dump was more of a concern.

In the early 1970s, Crow Wing County had built a central sanitary landfill, at the same time closing a number of smaller dumps throughout the area. One of them was quite near the Bocks' place—less than a quarter of a mile away. Dennis knew the site had merely been bulldozed over, that there had been no effort to clean it up or to remove anything toxic. Could something have leached out into the groundwater over the past quarter century that had now entered the lake? Dennis Bock spent the winter brooding about it.

———————

Hoppe, meanwhile, went back to Morris and reported what he'd seen to McKinnell and Helgen, who were stunned. McKinnell told me later the outbreak suddenly looked even more frightening when they learned that it was not limited to small agricutlural wetlands and manmade ponds. The researchers knew that if the Bocks' report were true it was evidence of deformed frogs in the kind of natural lakes that number in the thousands across the Minnesota countryside and are heavily used for recreation. This changed everything. "We were right back at the beginning," McKinnell said. Which was to say, more in the dark than ever.

Just as with Granite Falls two years back, the MPCA had conducted a few simple tests on sediment and tissue samples from the Ney farm and some of the other Henderson sites. The results had been inconclusive. They asked around to determine which were the major farm chemicals in use in the area. Reinitz and her students took periodic pH and temperature readings at the pond, too. None of this advanced the investigation, and by late fall—especially after the Bocks' situation had come to light—Helgen realized that if her agency and any of the other researchers were going to continue to work on the deformities problem next year they were going to need financing for a much more rigorous field program.

As it happened, the Minnesota state legislature was willing to oblige. The powerful State House Committee on the Environment and Natural Resources was then chaired by the late Willard Munger, a nearly eighty-year-old representative from Duluth, who as a kid earned money catching and selling frogs for bait. Munger knew what was happening at Henderson and was determined to do something about it. In October and again in February, Munger conducted hearings on the frog problem, and among those who testified were Helgen, McKinnell, and Reinitz and several of her students. At one of the meetings, Munger thundered that frogs were an important indicator of environmental health and that the problems turning up at Henderson and elsewhere needed to be taken seriously by everyone. "Whatever is happening to these frogs should be a warning to us that we have to do a better job of protecting our environment so there will be something left for us in the future when we all get down the line a little," Munger said.

In early 1996, with Munger's help, the Minnesota legislature approved $151,000 to study frogs. Hamline University in St. Paul received $28,000 of that amount to initiate an education program for elementary and secondary school students. The rest went to the MPCA. Helgen planned to continue the preliminary assaying of water, sediment, and tissue samples. She also contracted with McKinnell for dissection and analysis of frogs that they planned to collect in 1996 with Hoppe, who was to conduct population studies at several sites in an effort to determine just how significant the problem actually was in the field.

In retrospect, this didn't seem like much money or, frankly, much of an investigation. But as McKinnell explained to me later, the researchers were hesitant to launch a more ambitious—and expensive—effort given the uncertainties and their own lingering doubts about what they were seeing.

As the state huddled through the deep freeze of winter, Helgen, Mc-Kinnell, and Hoppe made their plans for the coming field season. Given their experience at Granite Falls, they were understandably nervous. Something had happened. The question now was, would it happen again next summer? As spring shouldered its way in behind the last snows and the ice thinned on lakes and wetlands across the state, the scientists fretted.

"Just as soon as we got that funding from the legislature we started to worry," McKinnell remembered later. "What if when we went out the next summer all we could find were normal frogs?"

But this time the outbreak of deformed frogs did not go away quietly over the winter. As the summer of 1996 unfolded, reports of deformed frogs poured in from every corner of Minnesota. One by one, the counties on Bob McKinnell's map were colored in. Hoppe cautioned that this didn't mean that every population of frogs was going to be affected. But at times it almost seemed that would be the case. Over the course of the season, the MPCA received 190 reports of deformed frogs—the majority were sightings of multiple abnormal animals—from citizens as well as officials at other government agencies.

The MPCA staggered under the crush of new findings. Helgen and her staff scrambled to investigate as many sightings as possible. They managed to confirm twenty-one sites with deformed animals in fourteen counties, from shallow, vernal wetlands in the southern corn belt to the sprawling marshes and lakes of the far north. The MPCA collected almost 3,000 juvenile leopard frogs over the summer. Just under 12 percent of the animals were deformed—although the incidence of deformities varied considerably from site to site and from one visit to the next. Some trips to affected sites turned up nothing but normal animals; other times there were double-digit percentages of abnormal ones. On July 25, only one out of 117 leopard frogs caught at the Ney farm was deformed. By September 30, 46 percent of the animals caught at Ney were abnormal. At Audre

Kramer's pond, just the reverse held. In July the rate of deformities was 65 percent. That fell to 36 percent in August and then to only 20 percent by early October.

Gleaning meaningful interpretations from this kind of data was close to impossible. The variables inherent in collecting frogs were so large that it was uncertain whether the trends were real or only apparent. Weather conditions, the skill and experience of the field collectors, and just plain luck all figured in how many frogs were caught and from which part of a wetland. These were factors that could shift dramatically in a single day at one site. But most sites were sampled once or at most a few times over the course of several months.

Even at that rate, what did the results of any of those collections actually indicate? Some animals were kept. Some were released after being examined. Few were marked so as to be recognized in subsequent collections. Maybe the same animals were being counted over and over again. The researchers also speculated that frogs crippled by limb deformities were easier to catch than the normal ones. After all, what was a field biologist but another kind of predator? Increased ease of capture would skew the ratios—though nobody had any idea how much.

But one observation held up consistently everywhere throughout the summer: Nobody found deformed adult frogs. All of the abnormalities were seen in young of the year—frogs that had just metamorphosed. Presumably, none of the deformed frogs seen in 1995 had survived. This wasn't altogether surprising—Helgen and McKinnell had duly noted the impact of the deformities on the frogs' locomotion. They fully expected that few of the frogs would be able to adequately feed and escape predators. Obviously, deformed animals must be dying off even as the scientists were trying to count them. So all that turned up in the data were the survivors. Who knew how many deformed frogs didn't live long enough to be counted?

Finally, the researchers had to consider the possibility that they were tracking only the most visible manifestation of a more complex syndrome, one that could have other, far-reaching systemic impacts on the animal. It was unknown whether the deformed amphibians had normal immune or nervous systems, or whether other organs were functioning properly, or whether they were diseased, or if there might be other internal structural deformities in addition to the malformed limbs. For that matter, no one could even say with certainty that frogs this year or the year before that looked normal really were normal. Perhaps not coincidentally, the MPCA had received three reports of declining frog populations that summer. The abundance of juvenile frogs at most of the sites the agency sampled suggested a relatively successful breeding season, but there was no long-term baseline data for comparison, so nobody could really say what impact the deformities might be having on local populations. Intuitively, it seemed unlikely that any wetland where a large percentage of offspring did not survive their first year could sustain normal population levels over time.

McKinnell began getting disturbing answers to some of these questions as soon as he started dissecting deformed frogs that summer. Because of his long experience in examining frogs for tumors, McKinnell had been assigned to perform necropsies in order to look for internal abnormalities. He expected to find some. Frogs with external limb malformations, he maintained, would be "remarkable if they were bereft of internal pathology."

McKinnell dissected more than five hundred frogs—including both normal and abnormal appearing ones—that were collected during the summer of 1996. And, just as he suspected, there were internal problems.

There was evidence of abnormal reproductive system development—a finding consistent with exposure to some external toxicant. Chemical agents that cause developmental abnormalities are called teratogens. Teratogens can impact many organ systems and body structures, but sex organs are a favorite target. McKinnell found testes that were dramatically undersized in male frogs, and because some were also virtually clear instead of opaque, as they normally are, he noted that in many cases they were hard to find even under the dissecting microscope.

Other abnormalities were found in digestive tracts. The most dramatic ones involved the stomach and lower gut, which in some of the animals were severely distended and stretched. McKinnell was astonished to see repeated instances in which the gut wall had become almost totally transparent, so that its contents were readily visible. Even the gut contents were unusual. Many of the frogs were literally full of undigested insects and insect parts. McKinnell also found undigested insect pieces in a number of bladders—an observation he found utterly incomprehensible until he determined that it was directly related to the swollen guts. The only explanation for what he was seeing, McKinnell realized, was that the frogs were unable to digest what they were eating. As more and more undigested food ballooned the gut, the frog could not expel this material. Frogs have only a single opening at the rear, a vent called the cloaca that provides a common outlet for urine, feces, sperm, and eggs. As undigested food blocked the cloaca, it began to back up and ultimately regressed into the bladder. McKinnell determined that despite being crammed to the bursting point with food, even the frogs with enough mobility to feed themselves were starving to death.

And there was one more thing. McKinnell knew that leopard frogs will only eat live food. They won't consume anything that doesn't move. Yet mixed in with the undigested insects, McKinnell also found small sticks, leaves, and even pebbles in a number of frog guts. While McKinnell conceded that he was at a loss as to what might explain this strange finding, he marked it down as potential evidence of a behavioral abnormality—something new to go along with the physical malformations.

Meanwhile, McKinnell and his assistants continued to marvel at the various physical abnormalities exhibited by frogs brought in from the field. Probably the best known, because Judy Helgen posted a photograph of it on the Internet, was one with a thin, pink, perfectly matched pair of extra legs growing out from between its normal hind legs. Another favorite was a leopard frog that had a very uncommon forelimb deformity—a well-developed extra leg that protruded from just behind the corner of frog's right jaw.

In late May of 1996, David Hoppe returned to the Bocks' lake, wondering all the way there if he'd find anything like what Dennis and Rhonda had described to him the previous fall. He arrived about ten o'clock in the morning and immediately heard frogs calling—not the earnest, heated chorus you'd expect to hear in the evening, but just the steady thrum of a lake alive with frogs. As he had anticipated after his first visit, Hoppe discovered a number of frog species in residence, including several that only an experienced field herpetologist would readily find. Listening carefully, Hoppe could hear several leopard frogs calling, plus a slightly larger contingent of American toads. Toads have one of the most delightful of all frog calls—a musical, high-pitched trill that at the height of the breeding season fills the summer night like nothing else. Hoppe was encouraged by the apparent abundance. Making his way down to the shore he heard a gray tree frog calling.

Hoppe waded into the reeds, edging through floating mats of vegetation. The air was heavy and close. The lake had a ripe, organic odor. Buzzing squadrons of insects swirled around him and oscillating shafts of sunlight angled through the leaves of the trees that hung out over the water. Hoppe moved carefully. It was slow, sweaty going. The bottom was a heavy muck, difficult but manageable for wading. Inside the vegetation line the water was only thigh-deep. He made his way slowly up the shore, pulling one foot out of the mud as the other sank in, inching his way toward the fence running down the bank and into the water at the edge of the property a hundred yards or so north of the dock.

The shoreline looked like frog heaven and it was. There were many frogs and a great many more large tadpoles submarining through the brown water as Hoppe pushed ahead. He caught sixteen adult frogs and two subadults, gently returning them to the water after a close inspection. They all looked fine. The lake seemed almost crowded with robust, healthy, normal amphibians.

Hoppe made a total of seven visits to Bocks' lake that summer. On land, he captured spring peepers, wood frogs, gray tree frogs, American toads. At the water's edge he netted leopard frogs. His major find was in the lake itself. Wading the shallows, he caught mink frogs, a highly aquatic species that immediately confirmed the lake was both a breeding and an overwintering site.

Adult mink frogs spend most of their time in the water, rarely venturing more than one or two jumps onto dry land. They are also slow developers. Unlike other frogs that metamorphose just weeks after hatching and can migrate from their breeding ponds, mink frogs typically spend one full year in the larval stage and pass their first winter as tadpoles in the same place they hatched. Mink frogs are attractive, sturdy animals—a tad smaller through the body than leopard frogs. They're a dark, mottled green in color, usually with a bright green chin.

Mink frogs get their name from their musky odor, which is more pronounced in adults. Although kids playing by lakes sometimes come across mink frogs, they are not nearly so commonly seen as other species. Hanging motionless in the water, with only their eyes and nostrils above the surface, or hunkered down in a tangle of aquatic vegetation, they can be almost impossible to spot from only a few feet away. Dennis and Rhonda said later that they were completely unaware of the mink frogs living in untold numbers just a few feet from their shoreline.

Hoppe had been initially excited by the extravagant numbers and the rich diversity of frogs he encountered. It was hard to believe there could be anything seriously wrong with the environment here. But if the deformities Dennis and Rhonda said they'd seen the year before were present again this year, Hoppe felt certain that whatever caused the abnormalities would be found in the lake, which seemed like it must be center stage for all of the frog species in the area. Hoppe stepped up his pace as the weather warmed, visiting the lake as often as he could during the peak of metamorphosis in order to see as many of the new crop of juvenile frogs as possible.

He was not prepared for what he found.

─────────────

The scene at the Bocks' lake deteriorated in front of Hoppe's eyes. By midsummer the lake had become a watery nightmare. Each successive collection revealed a fresh perversion of nature. The frogs reproducing along the Bocks' shoreline—where the family had swum and caught their Sunday supper for years, and where Hoppe had seen nothing out of the ordinary only weeks before—were being ravaged. There were significant levels of deformity in every species. The most severe abnormalities—the worst yet seen anywhere in Minnesota—occurred among the mink frogs.

The mink frogs at the Bocks' lake exhibited every type of deformity previously seen at other places in the state—plus a ghastly parade of new ones. Some were missing legs or leg parts, but most of the animals had some sort of twisted addition to their normal complement of limbs. Many had the strange skin webbings—Hoppe started calling them "cutaneous fusions"—that spanned the trailing edges of the hind legs, preventing their full extension and in the most extreme cases totally immobilizing the limbs. Hoppe found it hard to look at these animals and not wonder if they weren't likely to die by drowning.

Other frogs had stumps of legs or partly formed limbs protruding at odd angles from their hindquarters. There were legs that split, branching off in two or more directions, as well as legs that had extra feet, and feet that had extra toes. There were tapering, leglike appendages sticking out from the rear flank like wings on an airplane. One frog had a total of nine legs. Others were so badly deformed that Hoppe could not describe them in his field notes, or ran out of room trying. He took one of the worst cases— a mink frog with a confused tangle of hind limbs bunched in a fleshy mass and festooned with extra feet and toes too numerous to count—back to his lab in Morris. His students nicknamed the frog "Scrunch." Despite the most attentive care and regular hand-feeding, Scrunch died after about ten days in captivity.

As the summer deepened, Hoppe became convinced that the occurrence of such profound deformities and the presence of so many different species would provide new insights into what was happening. Hoppe had by then visited the Ney farm down in Henderson and had found a few toads with deformities, but the wide spectrum of species affected at the Bocks' lake was arresting.

All of the frogs here showed a range of limb abnormalities, though in general none seemed nearly so bad off as the mink frogs. What interested Hoppe were the varying incidences of deformities among the different species. Fully half of the mink frogs were deformed. About 10 percent of the leopard frogs were deformed. Among the wood frogs, spring peepers, gray tree frogs, and American toads the abnormalities occurred on average in only about 3 percent of the specimens he inspected—although this was still more than Hoppe had ever seen anywhere in all his years of field collecting.

Hoppe saw that if he arranged the frogs in order of their affinity for the water that the rates of deformity corresponded perfectly. The more aquatic the species, the higher the incidence of abnormalities. Surely this was yet more proof that a waterborne agent was implicated in the deformities. Or was it? The fact that adult mink frogs spend a lot of time in the water would help explain that species' higher susceptibility to deformities only if the malformations were the result of parental exposure to a toxicant in the lake—something that bioaccumulated in the tissues of the mother and was then passed on inside her eggs as a birth defect.

If the deformities did not result from such a maternal transfer—if they were developmental errors induced after conception—then the picture was murkier because there were fewer behavioral differences among the species. All of the frogs were in the water in the egg stage. And, of course, all were completely aquatic as tadpoles. They were immersed in the water throughout their development. On the other hand, they reproduced at different times and that could mean different exposures to a toxicant that was present in varying concentrations over the course of the season. In the end, though, Hoppe could not ignore the fact that mink frogs would endure a much longer exposure to anything in the water simply because they remained tadpoles throughout their first year.

Hoppe asked Dennis and Rhonda if he could invite the MPCA up to take samples and see the the appalling array of deformities in the lake.

They agreed. Dennis recalled later that they had decided they trusted Hoppe and that they would do whatever he recommended. They also got a visit from Bob McKinnell, who accompanied Hoppe one day. Dennis remembers asking McKinnell if he thought they should stop swimming in the lake. McKinnell said that looking at it objectively, he doubted very much that there was any risk in just going in the water. But he said he wouldn't do it.

"That kind of made an impression that stuck with us," Dennis told me. The Bocks stopped swimming in the lake.

Judy Helgen had been worried from the very beginning that landowners would be reluctant to cooperate with their investigation if word got out about where the deformities were being studied. Some would balk at possible attention from the press; others might be afraid the agency would accuse them of damaging the environment in some way. So the MPCA began assigning code-letter designations to their study sites. The Bocks' lake, which didn't have a name, was called CWB. As in "Crow Wing Bock." Dennis Bock told me one time that he and Rhonda had another, private name for it—a variation on nearby Nokay Lake. "We used to call it Nooky Lake," he said. He grinned shyly when he said it, but I thought he sounded a little wistful, too. By the end of the summer of 1996, even the Bocks had started calling the lake CWB.

CWB was the focus of Dave Hoppe's attention that summer. But it was not the only place he collected frogs, and it was not the only place where his collecting turned up surprising deformities. Hoppe found deformities at the first ten sites he went to that year and soon wondered if he would ever find a place where the animals were not deformed. He found significant levels of deformities at a number of sites where someone else had reported seeing a deformed animal or two, but also at some places chosen at random or where he and McKinnell had previously done field-work and had historical data. He found deformed frogs at sites not more than a few miles from the Morris campus, in wetlands where he'd often taken his classes and where he had personally handled thousands of frogs over the past twenty years without ever seeing a limb abnormality. One day he sent several of his students out to survey these wetlands. They found deformed frogs. While they were gone, Hoppe found a leopard frog with a missing leg in his own backyard.

The sudden appearance of deformed frogs seemingly everywhere created a problem for the scientists. Their research plan did not presume what the cause of the deformities might be. But it was tilted toward the possibility of a chemical toxicant. One important consideration, then, was to locate control sites where there were no deformities in an effort to identify chemical agents that were present in the affected sites but not in the controls. The researchers hoped to pair each study site with another nearby wetland that appeared similar in every respect, but where the incidence of limb abnormalities was less than 1 percent—preferably zero.

When the scientists began finding rates of deformity of 4 or 5 percent—or higher—at sites with no prior history of abnormalities, they realized that choosing wetlands to serve as control sites over a period of years might be next to impossible. Inevitably, this worry led to a question no one had yet seriously considered. What was the normal "background rate" of limb abnormalities in populations of wild frogs? Up until now, everyone had assumed that such deformities were extremely rare. After all, hadn't Hoppe and McKinnell, who'd held tens of thousands of Minnesota frogs in their hands over the years, said they'd never seen such things before?

The flood of reports from citizens around the state—and the accompanying flurry of local reports of deformed frogs in the papers—caused the researchers to ask themselves if the abnormalities were simply turning up now because for the first time a lot of people were out looking for them. Both Hoppe and McKinnell doubted that they would have failed to notice significant numbers of frogs with serious leg abnormalities, but they conceded that lesser deformities—say, the wrong number of toes, or a slightly undersized leg segment, or a mild instance of skin webbing—were things that could have escaped the attention of even the most seasoned field biologist who wasn't specifically looking for such deformities. Now those were exactly the sorts of abnormalities that were being seen and reported and counted. Plus, it seemed clear that there was only a short window of time to observe the deformities—a couple of months, generally—because none of the deformed animals survived their first year after metamorphosis. What if over the course of their many summers in the field, Hoppe and McKinnell had just never happened to be in the right place at the right time to learn that leg deformities are a relatively common occurrence in juvenile frogs?

This concern deepened as Hoppe began reviewing the scientific literature. Published findings are the lifeblood of science. Every scientist wants his or her work published in technical journals, where data and conclusions are subjected to the intense scrutiny of peer review. And every scientist relies on published findings in much the same way attorneys rely on the published precedents in case law. Published scientific work carries weight that mere observation does not. Scientists routinely debate the conclusions they reach, but the published data on which those ideas are based are generally recognized as fact until proven otherwise.

Hoppe quickly learned that deformed frogs were not at all a new phenomenon. In fact, they had been known to science for more than two centuries. There was even a published report from 1740 that was accompanied by drawings of multilegged frogs that looked uncannily like some of the deformed frogs that had turned up in Minnesota. Much more recently, in the 1980s, a California researcher had encountered a localized outbreak just south of San Francisco and had published the results of laboratory experiments that seemed to prove the deformities were caused by a common parasite.

It was hard for Hoppe to know what to make of the scientific record. It undeniably showed that limb abnormalities had occurred in frog populations before the Industrial Revolution and thus long before pollution from any modern chemicals could be involved. Yet the published reports seemed to have more in common with one another than with what Hoppe was seeing with his own eyes that summer.

In virtually all of the previously recorded outbreaks, extra legs were the only deformity noted. Hoppe had seen plenty of extra legs at CWB, but elsewhere across Minnesota extra legs were not the most common abnormalities. There seemed to be almost nothing in the literature about missing limbs or severely reduced limbs, and no mentions of skin webbings, either. It also appeared that oubreaks of deformities usually occurred in discrete populations, one species at a time. None of this paralleled what seemed to be happening in Minnesota. A notable exception was the California report linking limb deformities to parasitism, where a wide range of abnormalities, including all the ones seen in Minnesota, had been found in both tree frogs and salamanders in the same wetlands. But the experimental findings in that paper, which were quite limited, only suggested a link between parasites and extra legs, not the other deformities.

A number of the outbreaks previously recorded featured very high rates of the exact same kinds of leg duplications, so that all of the abnormal frogs usually looked roughly the same. This suggested to Hoppe the possibility that genetic mistakes might account for at least some portion of these reports. It could be that a single clutch of eggs from just one female frog from time to time resulted in a batches of multilegged juveniles. The offspring had sufficient locomotive difficulties that they did not reproduce. When they died off the deformities disappeared as suddenly as they had cropped up.

And even though there was a long record of periodic outbreaks, it wasn't as if they happened all the time. Outbreaks seemed to be isolated and short-lived, usually spanning only a single breeding season. In short, there was plenty to look at in the scientific literature on deformed frogs. But such episodes were evidently rare and not very much like what was going on in Minnesota. Even so, it was now clear that the researchers would need more data—a lot more—to be able to say with certainty that the Minnesota outbreak was outside the realm of normal biological mishap.

———————

The preliminary data, such as they were, already seemed to be creating a different sort of problem. Not yet a year old, the Minnesota frog investigation was showing signs of internal stress, especially when it came to processing information being gathered in the field. Back at MPCA headquarters in St. Paul, Judy Helgen was drowning in work. She and Mark Gernes, a wetlands botanist she'd been working with since before the frog problem showed up, had been traveling the state and looking at frogs all summer, only to come back to an office each time that was in worse disarray. They never expected the outbreak would mushroom as it had, or that it would come to take as much of their time as it did. Stacks of field data and citizen reports grew unchecked. Phone messages and e-mails backed up and went unanswered. Inquiries from local reporters wanting to know what was going on became a constant, nagging distraction. Helgen's supervisors—nonscientists for the most part—had meanwhile not seen fit to relieve her or Gernes of any of their other responsibilities. As the frog investigation pushed everything else aside, Helgen and Gernes fell seriously behind in their other work. Tempers began to fray all around.

Hoppe and McKinnell, who'd been eager at first to look into the frog situation, now found themselves increasingly uncomfortable with their roles in the investigation, which seemed ragged and less focused than they had hoped. Hoppe fumed over what he saw as needless "red tape" in accounting for his time and expenses. McKinnell bristled at the general lack of organization and the occasional suggestion that the MPCA was in charge of the inquiry. Both men looked askance at Helgen's travels from one wetland to another without a coherent idea of what she was looking for or how to evaluate what she found. Hoppe thought the situation at CWB demanded much more attention than it was getting. And both Hoppe and McKinnell worried that their accustomed independence as academic researchers was being compromised. They grew wary of what would become of their work once it went into the bureacratic maw in St. Paul. Privately, each of them decided they would share their findings with Helgen on a selective basis.

––––––––––

It would end up taking Helgen almost two years—until the spring of 1998—to produce a report from the harried field season of 1996. In the interim, everyone knew only in the most general way what was being learned and what was rumored to have been learned. Chemical analysis of frog tissues and wetland sediments performed by two private laboratories failed to turn up much of interest. Residues of various organic compounds were found, including some herbicides that Helgen was surprised to see accumulating in the tissues of young frogs. But these compounds were found in both normal and abnormal animals. There were slightly elevated levels of heavy metals at some sites where deformed frogs were found, but the amounts appeared to be well within accepted safe limits. Chromosomal analysis uncovered no relevant genetic abnormalities.

Examinations of frogs for parasite infection were given a high priority—Helgen and McKinnell sent out specimens to two different labs for evaluation. The results were inconclusive, but the preliminary evidence suggested no connection between parasites and limb defects. Most of the frogs examined had a pretty sizable load of parasites. This was expected, since parasites are a routine fact of life among most higher organisms. But not every frog was infected, including some of the abnormal ones. A deformed frog that didn't have parasites seemed a pretty strong argument against parasites as a cause of the abnormality. But Helgen wasn't completely persuaded by this that parasites could be ruled out, and she wouldn't be for a long time. She thought that shifts in the frequencies and types of deformities during the course of a single season might in some way correlate to changing populations of parasites, but that the ecology—what Dave Hoppe would call the biology of the situation—was extremely complicated. "It could be parasites," she told me fully a year later, in mid-1997. "I mean, it's possible. We want to keep an open mind. We want to stay open to all the possibilities. It's a mystery, you know."

Of course, mysteries deepen with the passage of time. The inability to solve a problem sharpens the perception that it is a difficult problem. In science, failure to find an easy answer at the outset of an investigation has its rewards, the most important of which is that it keeps the mystery intact and the money coming in. *We don't know what's wrong, but it appears to be worse than we thought.* After its first full year of work, the Minnesota deformed frog inquiry had achieved just such a result—the scientific equivalent of a ground-rule double.

Bob McKinnell, who intended to publish his findings on his own, provided Helgen with a written discussion of his dissections of abnormal guts and odd stomach contents—but not the actual data. This resulted in widespread reports that the Minnesota team had linked internal deformities to limb abnormalities. This proved to be a considerable overstatement.

When I spoke with him about these findings in late 1998, McKinnell—who by that time was retired and essentially out of the frog investigation—said he'd found internal abnormalities only in some animals from the Ney farm. They never turned up anywhere else and he had no evidence that they were correlated either with limb abnormalities or with wetlands that produced frogs with limb abnormalities, since he found them in both deformed frogs as well as frogs that appeared normal. McKinnell's report of internal problems was thus something of a dead end, an odd non sequitur in the MPCA report when it finally came out.

Hoppe's data were a different matter. The deformities seen in multiple species and the apparent correlation with aquatic life histories were electrifying bolts of new and troubling knowledge. But Hoppe also decided to hold back some of his findings. Hoppe didn't trust Helgen or the MPCA with his data, or with maintaining privacy at the sites where he expected to continue to do research. Giving the sites secret code letters wasn't enough.

In fact, Helgen shared Hoppe's concern about ensuring the scientific integrity of locations under study and with protecting the identities of a number of private landowners who were cooperating with the investigation. She worried about how their neighbors might regard them. Some people had already begun making inquiries about their safety. Other people—farmers in particular—who had found deformed frogs on their property were afraid they might somehow be held responsible for what was happening. Helgen didn't want public scrutiny of these people or their properties, which she was certain would only make the investigation more difficult. Don Ney's farm was already widely known and had been frequently visited—not only by the MPCA and Hoppe and McKinnell but by students and the media. Other locations—including CWB—remained closely kept secrets.

But since most of the work done by the MPCA is ultimately a matter of public record, Hoppe doubted that any information would be safe with the agency forever, regardless of Helgen's best intentions. And, like Mc-Kinnell, Hoppe felt he properly owned the discoveries he'd made that were beyond the scope of the work he'd been contracted to perform. In the end, Hoppe decided he was obliged to provide at most a general overview of his research and to submit hard data only on leopard frogs. This left out his more compelling work on mink frogs, but in the end it was a moot point. Hoppe's findings would never make it into the MPCA's final report on the 1996 field season, evidently because Helgen lost his data somewhere in the ocean of paperwork that had engulfed her.

So the investigation frayed and went on anyway, the researchers over-worked and increasingly puzzled by the deformities. Helgen prepared to go back to the Minnesota legislature for another round of funding for 1997, and Hoppe and McKinnell agreed to continue with the investigation despite their misgivings about how the MPCA was handling things.

Helgen was by now completely absorbed—obsessed, really—with the investigation. So was Dave Hoppe, who, perhaps more than anyone else, felt personally distressed by what he'd seen that summer. He'd gotten to know Dennis and Rhonda Bock and had frequently enlisted the Bock children in helping to collect and examine frogs. He thought often about the fact that the Bocks' drinking water came from a well within fifty yards of the lake. And he realized that the Bocks weren't just tolerating his frequent visits out of curiosity. They expected him to find out what had gone wrong in their lake.

As autumn pushed into Minnesota, Hoppe sat in his office at the university back in Morris and thought anxiously about the Bocks. Those searing days of summer when the animals he'd studied his whole life were being turned into grotesque parodies of frogs seemed remote, a little unreal. Now a chill wind stabbed across campus, roaring out of the Dakotas, bearing down on western Minnesota with news of a season to come. Hoppe had a long time to think about what he'd found.

Hoppe usually started off his fall field class in herpetology by surveying the students on what they planned to do after college. He'd pass out a questionnaire and, after a few minutes, collect it. Hoppe would entertain himself by reading silently through the responses and then, working to keep a straight face, announce to the class that he was very surprised but delighted to see that all but one of them hoped to become professional herpetologists. "They always look around the room really alarmed," he told me. "Everyone thinks they're the only normal one in a class full of weirdos."

Hoppe also liked to explain to his students the origin of the word *herpetology*. It comes from the Greek verb *herpein*, meaning to creep.

"Creepy crawly," Hoppe would add.

Herpetology. The filmy, dank science of slithery things living in dark, wet places. Maybe the last, best measure of how bad it had been in the wetlands of Minnesota that year was that the herpetologist himself now had the creeps. David Hoppe felt queasy just thinking about what he had seen. And he felt very alone. It was a strange time. Looking over the photographs of mink frogs he'd taken during the summer, Hoppe could not imagine that anyone else had ever seen anything like them before.

Someone had.

Surprised But Not Crazy

H ALF A CONTINENT AWAY, AT MCGILL UNIVERSITY IN MONTREAL, A young French-Canadian graduate student named Martin Ouellet knew exactly what Hoppe was going through. In July of 1992—three full years before Cindy Reinitz took her class on that fateful outing to the Ney farm— Ouellet had stood at the edge of a small pond in the middle of a field of corn not far from Montreal, holding a leopard frog with a strangely under-sized hind leg and wondering what to make of the animal. He was only twenty-three years old at the time. Before he reached thirty, Ouellet would see thousands more like it.

At first, Ouellet thought his discovery suggested that something had gone haywire in the environment. In time, he was convinced of it. This was not a conclusion that depended entirely on the evidence. Understanding what caused a frog to be deformed was one thing; understanding the symbolic implications was quite another. "Life is political," Ouellet told me once. We had been talking about the French-Canadian separatist movement, while hiking the wetlands of Mount St. Hillaire in the St. Lawrence River Valley. This was in the late summer of 1997, by which time Ouellet had examined more deformed frogs than anyone else in the world. Ouel-let, who was finicky and skeptical in his pursuit of scientific truth,

nonetheless saw a fit between the deformities problem and his environ-
mentalist learnings. When I asked him if he was for or against French
independence in Quebec, he just smiled. "I will never tell you," he said.
"And you will never guess."

He was more direct on the deformities question. From the moment
he'd found that first frog, Ouellet believed he'd discovered something that
would shake the public awake about environmental degradation, something of interest not just to scientists but to anyone who cared, or who
could be made to care, about the state of the planet. Like David Hoppe, he
was soon persuaded that some deformities could arise naturally. They'd
been reported for centuries and presumably occurred from time to time for
as long as frogs had existed. But he was also convinced that something
new and different was happening on a wider scale. Once, when I asked
him about parasites and deformities he thought for a moment and then
answered that such a connection probably existed, but it really didn't matter. By then, Ouellet had come to believe that frog deformities have many
causes, some natural and some quite unnatural.

"Just about anything can cause deformities," he said. Ouellet speaks
wonderful, precise English that he considers and draws out syllable by syllable. He pronounced the word as *day-form-a-tays*. "High temperatures can
cause deformities," he continued. "Ultraviolet light can cause deformities.
Chemicals can cause deformities. A lot of things. I think, maybe, Diet Pepsi
can do it, you know? I know parasites can do it. I'm sure of that. But it
won't explain everything. Parasites are not the primary cause. We will
prove that. It's clear-cut. Myself, I don't care what is causing deformities.
My worry is that if people think that it is only parasites then they will continue to pollute."

There's an old saying in science that you usually find only the data
you are looking for. Martin Ouellet walked into a cornfield in 1992 looking
for evidence that humans are making the world an unfit place for frogs. He
walked out with a discovery that changed his life. Not everyone would see
it quite the way he did. But Ouellet was a confident young man who
understood what he'd found.

"It's clear-cut," he told me. "Totally."

Ouellet grew up in Québec City, one of some 800,000 intensely independent French inhabitants. The city has an Old World flavor, and Ouellet
comes across that way, too. He's trim and handsome in a wire-rimmed-

intellectual way. As a youngster, he divided his time between the city and a cottage his parents owned. His mother, an amateur naturalist, indulged her son's avid interest in frogs and snakes. "Our place was on a small lake in the country," he said. "In the middle of nowhere. It was having plenty of wildlife. When I was young I was fishing. I was hunting. Until I realized that I was killing and I was not having any good feeling doing that. So now I am basically against fishing and hunting. Now I have no pleasure anymore to do that."

Following his inclination, Ouellet attended the only French-speaking school of veterinary medicine in North America, at the University of Montreal's campus in St. Hyacinthe. He earned his degree and began practice as a vet in 1991. His specialty was exotic animals: parrots, reptiles, even the occasional frog. It was not an altogether happy experience. Ouellet found himself increasingly disturbed by the keeping of certain kinds of animals as pets. He disliked cages and found it unpleasant to treat animals he thought should be living in the wild. At about the same time, the Canadian Wildlife Service was planning a project to assess the impact of pesticides on nontargeted wildlife. The agency had already tried monitoring the effects of modern agricultural practices on birds. But the project ran into problems. Birds were difficult to work on. They moved around a lot and were hard to capture. It turned out, too, that birds were as readily disturbed in the wild by biologists as they were by farmers. So the agency decided to reconfigure their research around a different class of animals: amphibians.

Amphibians live and breed in large numbers in agricultural areas. Their movements are more restricted than birds and they are easier to catch. But there was another reason the agency got interested in amphibians. In recent years, scientists around the world had begun to notice apparent declines in amphibian populations—especially among frogs—that could not be easily explained as the result of human encroachment and habitat loss. Biologists had been trading anecdotal reports of frog declines and disappearances for some time, but their concerns had lately been more formally acknowledged at the First World Congress of Herpetology in Canterbury, England, in 1989, and again at a meeting that was held in Irvine, California, in February 1990. Many theories had been advanced to account for the declines, including acid rain, climate change, new diseases, increases in ultraviolet radiation associated with ozone depletion, and contamination from manmade chemicals, including pesticides. All the wildlife service needed was someone with clinical skills to work on amphibians—mainly field examinations and the taking of blood samples. Presently, they learned about Ouellet. "They knew I was sort of crazy about reptiles," Ouellet recalled. "So they asked if I would try something with frogs."

In 1992 Ouellet had lined up a summer position at the Quebec Zoo, where he was going to work on an artificial insemination project for endangered species. The job came with a two-week vacation. He decided he could spend it looking at frogs.

Since then, Ouellet had examined thousands of deformed frogs, all of them caught in the heavily agricultural St. Lawrence River valley in Quebec province, between Montreal and Québec City. By 1996 Ouellet's cavernous room in McGill's fortresslike Redpath Museum was lined with filing cabinets holding the grim record of what he'd discovered—field notes and photographs that detailed every kind of limb abnormality imaginable, a veritable encyclopedia of warped morphologies. Missing legs. Extra legs. Reduced legs. Twisted legs. Legs that seemed to have been emptied of their bones. Skin webbings. Bizarre variations on all of these deformities that defied description.

Everything Hoppe had seen at the Bocks' lake Ouellet had seen many times over, at many sites, and during the course of several seasons. For Ouellet, the frog problem had become a full-time job, a personal crusade. He wanted to know everything about deformed frogs. And he felt that he was getting close. He knew what had been written. He knew what other scientists had discovered and that he'd discovered more than anyone else. He knew his way in the field—where to find frogs, how to catch them. He'd handled so many deformed frogs and become so familiar with the way they looked and felt that he could examine hundreds of animals in a single day. He knew what their blood looked like under a microscrope and what sorts of abnormalities to expect in their internal organs. He knew something horrible was wrong. He was quietly sure he knew one other thing, too. He believed he knew what caused the deformities. It was something he'd known intuitively right from the start. He was busy now trying to prove it.

Ouellet had seen pictures of Minnesota's deformed frogs on the Internet and was keenly interested in what was happening in the States. He wondered how the investigation was going. But he puzzled over the level of concern that had apparently developed there in the absence of any kind of broad field survey to assess the scope of the problem. The Americans, it appeared to him, were reacting reflexively before gathering up the necessary evidence. Mostly, though, Ouellet was glad for the company, even though nobody in the States seemed to be aware of who he was or what he was doing. For a long time he'd felt burdened by his knowledge of an environmental calamity. "I was by myself," Ouellet told me. "I was having nobody to talk with. I was feeling alone in this world."

The Canadian Wildlife Service set out to examine frogs living in agricultural settings where pesticides were in use and to compare them with frogs living in non-farm, control habitats that were similar but presumably free of pesticides. The study goals were modest. The agency wanted to get an overall picture of the general health and status of amphibians in the region and learn which of several clinical methods worked best at detecting the effects of exposure to sublethal doses of pesticides. The researchers hoped to do several kinds of blood tests on a large sample of frogs and take a much smaller number of animals back to the lab for dissection to look for any internal pathologies. They expected to find some. It was already well established that fish in the St. Lawrence River had unusually high incidences of cancerous tumors. So did the beluga whales that swam in the same waters. Ouellet thought it was likely he'd see similar kinds of tissue abnormalities in frogs. "I was expecting to find that a lot," he said. "Cancer. I was not expecting to find deformities."

But he did. The pond at that first cornfield Ouellet visited near Montreal, which the researchers listed as VE32 in their site codes, was a fairly typical small wetland. It was less than fifty yards across, maybe ten feet deep at its deepest, and surrounded by cropland. In just the few hours Ouellet spent there in the summer of 1992, he managed to catch sixty-one leopard frogs. Forty of them—66 percent—had hind limb abnormalities. He also caught fifty-four American toads. Thirty-seven of them—69 percent—were also deformed. Ouellet found one green frog, a highly aquatic species with a life history similar to a mink frog. It was deformed.

When Ouellet examined the first deformed leopard frog he assumed its stunted hind leg was merely the sort of unusual freak occurrence that turns up now and then in nature. Later in the day, as the numbers added up numbingly, the picture became more troubling. But it was only the beginning. Over the next two weeks, Ouellet found significant numbers of deformed frogs at five more agricultural sites. Meanwhile, his collections at control sites turned up only two animals with limb irregularities. What

Ouellet had at first assumed was a rare find at VE32 suddenly seemed more like the norm wherever pesiticides were in use.

"At VE32 I was surprised but not crazy," Ouellet told me. "But when we finished I was becoming a bit more crazy. Deformities were everywhere we were going that was subject to pesticides. So I started to think, maybe it's indicative of something, you know?" He said, *in-dee-kay-teev*. "So from there we began to think that this is maybe a good way to assess whether there's contamination," he went on. "It's maybe true that frogs are a good indicator. These deformities are basically good indicators that something is bad in the wild."

Something is bad in the wild. Real science is about inference and interpretation, about formulating hypotheses that can withstand every test to which they are put. To make anything from your data you need a skeptical soul. The only thing a good scientist can ever prove is that he or she is wrong. This is at the heart of the scientific method. Say what you think is happening and then try to disprove it every way you can. If you cannot, then just maybe you've got it right. Ouellet, despite an undeniable youthful exuberance and a heavy presumption that pesticides were bad for the environment, never assumed that he knew what he could not have known. From the start, he wanted proof. And he was not unaware of the irony in what happened to him on that day in July. In only a few hours on the job as a novice field biologist, Ouellet had happened upon solid evidence of an environmental crisis. In the arid prose of the paper he eventually published on his work in that summer of 1992 and the following year, Ouellet described part of what he did at the site: "[Frogs] were examined alive immediately after capture and any macroscopic abnormalities were noted." It was the first time he'd ever worked on frogs in the field.

Although Ouellet regarded his 1992 samples as too small—a total of 727 metamorphosing frogs and toads from nineteen sites were examined—the results, especially from VE32, were so chilling that there was no question that the research would continue on in 1993, with a new emphasis on limb deformities and what they might mean. The next year they collected nearly four hundred frogs from VE32 and seven new sites. At the end of the second season the overall rate of hind limb deformities in sites exposed to pesticides—even including several sites where no abnormalities were found—was 12 percent.

Ouellet and several colleagues completed a battery of blood and tis-

sue examinations, necropsies, DNA profiles, and toxicity tests in conjunc-
tion with their fieldwork in 1992 and 1993. They found that common
diseases and injuries occurred with pretty much equal frequencies everywhere. Some tests did suggest a toxic action by pesticides in the agricultural sites. One was a set of genotoxicity assays. DNA alterations were
monitored in laboratory cell cultures that were exposed to samples of
water from the ponds. Water from agriculture sites turned out to have
genotoxic effects at a level, the researchers reported, "comparable with
industrial effluents of intermediate toxicity." One pond where the water
proved highly toxic was a site Ouellet considered to be among the most
contaminated in his survey. It sat at the foot of a hill below a cultivated
field. On one occasion after it rained, Ouellet had seen a white foam form
on the surface of the water there.

Ouellet explored the toxicity question further in a series of caged
rearing tests, in which tadpoles were confined in enclosures in the pond at
VE32. The experiments didn't produce any abnormal limbs, but there was
a high rate of mortality—nearly 80 percent of the animals died in one
cage—and a noticeable delay in metamorphosis. Another test, which had
never before been used on wild-caught amphibians, involved a technique
called flow cytometry, in which blood samples are projected in a fluid
stream through a detector that excites and then profiles DNA. Exposure to
toxicants can cause changes in the genome—the total genetic complement
of the organism—often as the result of mistakes that are caused during cell
division, when the DNA unwinds then replicates itself in the process called
mitosis. Flow cytometry measurements showed abnormal DNA profiles in
frogs taken from agricultural settings.

These findings dovetailed neatly. Here was evidence that water in the
agricultural ponds could perturb DNA. Here also was evidence that DNA in
frogs from those same ponds had in fact been altered. And while it seemed
unlikely that the limb abnormalities were the result of DNA alterations,
the coexistence of both phenomena at many of the same sites could not be
ignored. Most probably both DNA damage and limb deformities resulted
from a common environmental insult—possibly one that was also lethal to
large numbers of frog larvae. Taken together, the results seemed to add up.

But this rising tide of circumstantial evidence was far from conclusive. As often happens in science, when the researchers looked carefully at
their data they saw that the facts gathered thus far told them more about
what they didn't know than what they did. In spite of what appeared to be
inescapable correlations, every finding seemed to open up a corresponding

hole in the tentative conclusions. One problem was that the blood work had to be performed on adult animals because the juveniles were too small to get samples from. Just as the researchers later found in Minnesota, all of the limb abnormalities Ouellet recorded occurred in newly meta-morphosed juvenile frogs—and in some case, maturing tadpoles. The Canadians guessed, as the Americans did too, that the limb defects were maladaptive. They impeded the frogs' ability to escape predators or to for-age successfully, and the animals never lived to maturity. So the value of blood work done on adult frogs was questionable.

A bigger deficiency in the data was the size of Ouellet's sample. At the end of the 1993 field season, he'd examined a total of 1,124 metamor-phosing frogs at twenty-six sites. But some sites yielded such small num-bers of frogs, often only a handful of a given species, that the data were not statistically significant. At VE32, for example, Ouellet found double-digit rates of deformity in two species in 1992, but in neither collection did he have even a hundred frogs. Meanwhile, an occasional deformity would show up in a "control" site, throwing those numbers out of whack.

The extreme variability, including data from the control sites, only increased when numbers from different collecting seasons were compared. Nothing really added up, and although Ouellet was eventually able to pub-lish his findings the study was statistically a flop—even though Ouellet felt sure that the control site deformities were flukes.

But Ouellet had learned a valuable lesson. In the coming years increasing the sample size would become his primary objective. The more frogs Ouellet could examine, the more he could say with certainty. By the time the Minnesotans began picking up frogs with deformities, Ouellet's data was much stronger—as was his conviction that pesticides were caus-ing the deformities. But he remained obsessed with getting larger and still larger numbers of frogs from more and more sites. He got very good at it, but was occasionally reminded that sampling populations of wild animals is never easy or foolproof. For instance, when Ouellet revisited VE32 a third time, in 1994, he could not find a single frog. Given the deformities he'd seen there the year before, Ouellet couldn't help but speculate on a connection. What he did see literally, as he would later write, was that the pond was by then "visibly polluted by a rustlike foam of unknown origin."

Scant as his initial data were, Ouellet concluded that nothing like it had ever been seen before. Like the others who would come into the investigation later, he searched the scientific literature for papers on limb deformities in amphibians. He found a lot that other researchers would eventually miss, because it turned out that the world's leading authority on deformed frogs was a French scientist, the late Jean Rostand. Between 1947 and 1970, Rostand had published an astonishing total of thirty-eight papers on frog abnormalities under titles such as *Les anomalies des amphibiens anoures* and *Les étangs à monstres: Histoire d'une recherche*. None of Rostand's work was ever translated into English.

"I came across it very quickly," Ouellet told me. "I realized there was already some stuff done. That guy, Jean Rostand, nobody talks about him now. He studied frog deformities for thirty-five years."

But in thirty-five years of work Rostand couldn't claim to have seen what Ouellet had found at VE32 and the other sites in Quebec. And certainly never on such a large scale. He had not seen significant occurrences of missing limbs. None of the other scientific papers ever talked about missing limbs, either. These deformities were strikingly not in the literature. But they were the *only* kinds of deformities that Ouellet had found. Missing legs. Reduced legs. Missing and reduced digits. Later he would find everything, the whole gory array of deformities. But in those first two seasons there had not been a single extra leg, not even an extra toe.

Ouellet did find one paper—from Russia—that discussed an outbreak of frogs with missing legs and missing toes, conditions that are known scientifically as ectromelia and ectrodactyly. But in that case the frogs had come from waters directly contaminated by sewage and city storm drains and exhibited not only missing limbs but a whole slew of cancers and other problems. It would take more than three years before Ouellet's findings from 1992 and 1993 were published, but when they finally were he noted that this represented the first documentation of a mass outbreak of ectromelia and ectrodactyly in frogs ever recorded in North America.

Of course, a second outbreak was, by the time Ouellet's paper was in press, well under way in Minnestoa. While researchers in the States were looking in vain for some common denominator among the sites where deformed frogs had been found, Ouellet continued to mount a case against pesticides. With each passing year, the data became more convincing. But there was still so far to go, so much work to do to tease out the precise compounds and biological mechanisms involved. The difficulties were especially hard to explain to the public—nonscientists—who were beginning to take an interest in reports of deformed frogs and who assumed that modern science could pretty much tell what the chemical constituency of anything was. Why couldn't scientists simply collect some water, take it back to the lab, break it down to see what was in it that shouldn't be there, and then check to see if the same compound was found at all the places where there were deformed frogs?

Because it does not work that way. Because analytical chemistry is a finite science and the possibilities were almost infinite. Because, as Ouellet often put it, *"It's dee 'el."* The hell.

Ouellet recognized that no matter how large his sample size grew, he could never prove conclusively that pesticides were the cause of the deformities until he got beyond simple guilt by association. The Canadian researchers—who unlike their American counterparts seemed to have a kind of quasi-police status that gave them access to private land—knew from interviews with farmers which pesticides were being applied to fields in the areas they were studying, as well as how often. But they could only speculate that those compounds were in fact getting into the pond water and that they were ultimately responsible for producing the limb abnormalities. They could not demonstrate this. They did not know what was in the pond water—including any compounds other than pesticides that might be implicated in the deformities—nor did they have direct evidence of pesticide contamination in the frogs themselves. And what they did know only made figuring any of this out seem an even more daunting task.

They learned, for example, that at least thirteen different herbicides were being applied as many as three times a year on all of the croplands. Some fields were also treated with any of at least five insecticides. Still others got hit with herbicides, insecticides, and one of a couple of commonly used fungicides. The mixing and matching of different products, and the varying times and frequency of the applications, created a congested intersection of potential exposures and dose levels. Pond water in agricultural settings, Ouellet came to believe, was "chemical soup" as he called it. Who could possibly tell what these frogs were being exposed to? Or when? Or how much?

Ouellet also knew that things were further complicated because of the many ways that pesticides might get into the ponds. They could simply fall into the wetland if there was enough wind to blow them over the water when they were being applied, or if the farmer were careless about getting too close to the shoreline. And certainly some pesticides would wash into the ponds in rainwater runoff. It seemed less likely, perhaps, but certainly possible, that pesticides could seep into the earth and be transmitted through groundwater into the ponds. Ouellet also saw undeniable evidence that in some cases, at least, pesticides were directly dumped in the ponds. He knew that some farmers used pond water for mixing pesticide formulations that called for water. Some farmers even used pond water to rinse out their tanks. In some places the evidence of contamination was too blatant to miss. Time and again, Ouellet found empty pesticide containers littering the banks or submerged in the waters of agricultural ponds.

More troubling still, the chemical soup was often in constant circulation. Some ponds were used as sources for irrigation water that was pumped out and sprayed over the fields, thereby creating a closed cycle of contamination and recontamination. Pollutants concentrated in the wetlands were respread across the landscape, and then recollected again in the pond—over and over again ceaselessly. Three sprayings of pesticides during a season could thus turn into a never-ending stream of continuing contamination. As Ouellet's survey expanded and he began looking at more kinds of croplands, this scenario took on even darker portents. Some of his study sites were adjacent to crops like raspberries and strawberries that were trucked directly to market or even sold from roadside stands at the farms. It was one thing for people to be unconcerned about pesticide residues that might be on a ripe red strawberry purchased right at the source. These compounds, after all, were approved for that use. But what would someone think if they knew that same strawberry had been soaked the night before with water from a pond crawling with deformed frogs?

So why just not test each of these pesticide products for toxicity? With enough time and money—Ouellet had the former but not very much of the latter at that point—you could do that. But what would it show? The fact was that science cannot deconstruct the components of such a chemical soup in the same way a gourmet can discern the ingredients in an elaborate dish. Any attempt even to start in that direction raised a lot of questions. What if limb abnormalities could only result from some transformation or activation that occurred in a natural setting that was too complex to model in the lab? What would be the dose? What would be the route of exposure? Were frogs affected because they swam as tadpoles in pesticide-laced water—or did they have to consume something from the food chain that picked up an increasing pesticide load over time? Such bioaccumulation of toxins had been seen in the food sources of other species.

Maybe the guilty compound had to be activated by sunlight before it could cause abnormalities. Maybe it was switched on by the natural water chemistry of the pond itself. Maybe it was hydrophobic and wasn't in the water, but rather the sediments or aquatic plants. Or maybe two or more compounds were interacting in a synergistic way, adding or even multiplying each other's effects, or even creating a joint toxicity out of compounds that were biologically harmless on their own. This is rare—any scientist who discovered such a bioactive synergism would probably patent it. But it was possible. Testing wasn't a bad idea—maybe you would get lucky. Maybe not. Maybes multiplied maybes.

This was "the hell."

And the hell had a bottomless, intractable quality to it. Ouellet's chemical soup was a black hole of uncertainty for still one more reason—something that would confound everyone who would eventually come to work on the frog problem. There was no assurance—it was in fact highly unlikely—that whatever compound the researchers were looking for was actually any of the pesticides that were being used. More likely it was another chemical form altogether—what scientists call a "daughter compound." An offspring, so to speak. That's because pesticides tend to break down rapidly in the environment. Many are specifically formulated to degrade in this way, the assumption being that it's a good thing for these toxins to go away as soon as they've done their job.

Pesticide manufacturers report the rate of breakdown of their products when they register them for use. But they are not required to tell—or even to know, for that matter—what these bioactive compounds break down into. Pesticides are tested for "safety," but whatever they leave behind in the environment after they disappear is not. And who could tell what that might be? How toxic were these mystery daughter compounds? How persistent were they? Did they break down quickly, too? Or did they linger for a long time? Did they in turn leave behind still more daughter compounds to worry about? How many? Even if you could isolate and identify the toxin you were looking for, how would you figure out where it came from? Every pesticide could spawn generations of different compounds. The chemical soup was more like a chemical orphanage, filled with organic offspring of unknown lineage.

This was the dilemma Martin Ouellet found himself in after a pair of two-week summer vacations in the fields of Quebec Province. "We were finding strange and weird things," he told me. "That stimulated my interest. I was more interested in this project," he said. "So I skipped my clinic."

"What do you mean?" I said.

"I quit," Ouellet said. In 1994 he enrolled as a Ph.D. candidate at McGill University so he could devote himself to the frog problem. His adviser, David Green, was one of Canada's leading herpetologists and head of the country's endangered species program. Ouellet set out to turn the wildlife service's pilot study into something much more substantial. "We had a low budget, basically," he said to me. "At the time our goal was to catch a couple of frogs, to do blood sampling, you know? We were trying methodologies. We were not knowing even if it was possible to find frogs in a pond in the middle of a cornfield. Nobody was doing that, you know? So it was highly exploratory. To see what is good, what is bad, what is possible, what is impossible. But we found more and more deformities. So we decided to investigate."

"That's the story about that," he added.

5
Metamorphosis

TRANFORMATION IS AN INHERENT PART OF LIFE. FOR FROGS, IT'S their signature act. But as swift and dramatic as the process of metamorphosis is—a leopard frog makes the jump from tadpole to frog in as little as three days—it is more notable for being so predictable. Seasons shift, landscapes wander, but time returns successive generations in all places as before. It's an annual cycle millions of years in the making, and each time the process is refined ever so slightly as reproduction begets evolution and vice versa. The genetic profile of an organism is its *genotype*. The mature, final version of the organism is its *phenotype*. Nature selects those slight, incremental variations between generations that produce better-adapted phenotypes and perpetuates them. Yet the genetic blueprint for every organism remains essentially stable. A young animal should look like a ripening version of its parents.

Clearly, something had gone terribly wrong at Don Ney's farm pond in that brief hot summer of 1995. It happened again the next year, at the Ney pond and seemingly everywhere across Minnesota. The same thing, or something like it, seemed to be going on in Canada. Somehow an annual cycle repeated over countless millennia had suddenly misfired. In the narrow band of time from spring to fall whole populations of animals were

taking a wrong step somewhere between genotype and phenotype. Intuitively, Cindy Reinitz's students had reached certain conclusions in the
space of a single afternoon that it would take many scientists many more
months to begin to believe.

The students assumed, for starters, that deformed frogs are not a normal occurrence. This was to be fiercely debated. The students also guessed
that something in the water had caused the deformities. This, too, was to
become a subject of the most intense inquiry. Finally, the kids speculated
that if something in the local water could do this to frogs, it could very
likely have harmful effects on other kinds of animals, including people.

———————

Judy Helgen had begun thinking about these and many other questions
the instant she'd gotten off the phone with Joel Chirhart. In much the
same way that Chirhart had been sent to the Ney farm just because he
happened to be available at the time, Helgen had now been tapped for a
job largely because there was no one else around to do it.

In the course of her Ph.D. studies at the University of Minnesota Helgen worked at the Itasca Field Station in the northern part of the state,
where she became intrigued with aquatic invertebrates, an unglamourous
class of organisms that includes snails, leeches, and crustaceans, plus the
many forms of insects that live in wetlands during their larval stages.
Whatever they may lack in charisma invertebrates make up for in their
teeming numbers and in their intimacy with the natural environment,
characteristics that make invertebrate biota potentially important ecological indicators. In 1989, Helgen had landed a job at the Minnesota Pollution
Control Agency where she joined the water quality section. It was a temporary, unclassified position, which Helgen had to fund herself with "soft
money" from outside sources.

Helgen brought with her to the MPCA a grant from the Minnesota
legislature and subsequently received additional funding from the EPA to
develop an invertebrate index, which she hoped would begin to correlate
species diversity and health with exposure to various environmental
stresses. When the 1993 report of deformed frogs near Granite Falls came
in, the MPCA didn't have a staff herpetologist. They did have Judy Helgen,
who had waders and knew how to use them.

Two years later, Granite Falls remained a puzzle. Whatever had
caused the frogs there to be deformed had apparently come and gone. But
Helgen suspected that if the cause had been chemical contamination of

some kind, there was probably an uncomplicated explanation for it. In 1993, weeks of heavy rain throughout the spring had caused extensive flooding across the upper Midwest, including the Minnesota River basin. The resident who reported the abnormal frogs had found them that summer around a small wetland in her backyard—an area that had been overspread by floodwaters. Helgen thought it was almost certain that the rising river had redeposited a potent mix of pollutants over the landscape and that those contaminants could easily have included compounds that could have caused developmental abnormalities. Among the possible sources of contamination in the area were an abandoned transformer salvage operation, an old city dump, ash ponds at a nearby coal-fired power plant, and the filthy Minnesota River itself. Given the many possibilities—plus the fact that no abnormal frogs were found the following spring—it had been hard to know whether to worry about this strange episode or just forget the whole business. The discovery at the Ney farm had brought it all back.

When Helgen and McKinnell visited the Ney farm they had tried to form a coherent picture of life there as the frogs knew it. As suspicious as the wetland looked—McKinnell shared Helgen's concern that digging might have raised something toxic out of the soil and into the water—he also knew that the frogs were being exposed to more than whatever might be in Don Ney's pond. They had found the frogs at a critical juncture. But leopard frogs are typically migratory animals with a multiphasic life cycle. They interact with the environment in many ways and in many places.

Frogs begin life literally immersed in their environment, and the interplay between a frog and its surroundings changes but never diminishes. While most other vertebrates develop within the protective confines of a womb or an egg, much of a frog's development occurs in the water at a larval stage. Different frog species have different natural histories. The story of the leopard frog is fairly typical.

It all starts in the spring. As wetlands thaw in warming weather, male leopard frogs initiate the breeding season by beginning to call to females. The call is a guttural, snorelike *yuuwwahh*, a sort of elongated version of the sound your hand makes if you pull it firmly across the surface of a child's rubber balloon. Female leopard frogs find this studly vocalizing irresistible. There is even evidence that they show a preference for lower-voiced males, and it has been observed that the males, whose calls drop in timber in warmer water, will position themselves near places where warmer runoffs enter a pond so as to sound more attractive.

Not that mutual consent is much of an issue once the raucous spring chorus of croaking and bumping and grinding has begun in earnest. Male leopard frogs are ferocious lovers, grasping their mates from behind and above, hooking their strong, oversized thumbs underneath their partners to gain better purchase. The eggs are fertilized as they are laid in the water during this embrace, which is called amplexus. Males turn out to be not too particular about the whole business. They will readily mount other males—who hastily identify themselves as such with a special "release call." They have also been known to attempt mating with inanimate objects. One Minnesota herpetologist reported a springtime conjugal visit between a male *Rana pipiens* and a floating beer can.

Leopard frogs breed in shallow, often temporary ponds, usually between April and June, with each female depositing several thousand eggs in a fist-sized, gelatinous glob attached to the underwater stem of a aquatic plant. The eggs and the jelly mass enveloping them are semi-permeable. Dissolved oxygen readily passes into the eggs, as does water. After about ten days, the eggs hatch and the embryonic frogs become free-swimming larvae—tadpoles. During this stage the tapoles are mainly her-bivorous, eating a variety of pond plants, especially algae. But they tend to be opportunistic feeders and will consume pretty much anything they can swallow. Under certain conditions, tadpoles can be cannibalistic.

Tadpoles breathe via internal gills, pumping water in through their mouths and out a small, tubelike opening on the left side of the body. In shallow wetlands, especially vernal ponds that dry out over the course of the summer, oxygen levels may drop so low that the tadpoles must come to the surface and gulp air to breathe. This low-oxygen condition is called hypoxia, and it often occurs cyclically at night, when photosynthesis stops and aquatic plants go into reverse, taking up oxygen from the water and expelling carbon dioxide—the opposite of what happens in the daylight hours. Tadpoles are also highly responsive to temperature. Warmer condi-tions generally speed development, which is fortunate because hot weather also tends to dry out wetlands faster. As midsummer approaches and tadpoles reach metamorphic climax, the animals are sometimes racing against time to get out of the water before it disappears altogether.

Metamorphosis is swift. The tiny limbs that have formed on either side of the tail suddenly grow into powerful hind legs. Front legs, which have developed internally within the branchial chambers, beneath the skin of the chest, emerge, breaking out out through the skin fully formed. The pace accelerates dramatically as other changes begin to take place. Skull and skeletal structures are radically reconfigured. Internal organs are transformed. Gills disappear and lungs develop. The animals become increasingly carnivorous. Finally, the tail degenerates and is resorbed, fueling the final conversion to frog. Once on land, juvenile leopard frogs disperse in search of food. Sometimes they don't get far. Adult frogs, always on the lookout for a hearty meal, have been known to return to the edges of their breeding ponds to feed on the juveniles as they metamorphose.

Even away from the water, frogs continue to bathe in and soak up their surroundings. After metamorphosis, frog skin remains quite permeable. Leopard frogs as well as most other species of frog constantly take up both water and oxygen through their skin. The extent to which their skin permeability increases exposure to environmental factors is a somewhat contentious subject among scientists. But there is broad agreement that the physical boundary between the frogs and their surroundings might be easily crossed.

On land, frogs supplement lung breathing with skin respiration, and this requires that their skin stay moist. Dry conditions are hard on frogs, whose bodies are about three-fourths water. But frogs generally drink very little orally. Instead, water is freely absorbed through the skin. Most frogs on land absorb water right off the surface of the ground, usually through thin, blood-enriched areas of skin on the underside of their torso between the hind legs. The American toad, for instance, does its drinking through a translucent patch of skin beneath its rump called the "sit spot."

As summer continues frogs move farther from the water. In northern climes, this phase is short; in fall the frogs migrate en masse across the countryside, heading back to the water. Amphibians living in cold climates have evolved a number of ways—some quite amazing—to survive winter. The American toad burrows underground and moves higher or lower with the frost line, staying just below the layer of frozen earth. The wood frog, a small, uncommonly handsome frog of the uplands, spends its winters on the ground beneath leaf litter, nestled among tree roots, or lying under rocks. With its dull brown skin and a sharply outlined black "mask" across its face, the animal is pretty well camouflaged in the forest. But it ranges north to the tundra above the Arctic Circle, and there are isolated reports in Minnesota of wood frogs jumping about over the snow in January. The wood frog's secret: it can survive partial freezing. In winter the wood frog's body becomes stiff and its eyes glaze over. As ice begins to form in its tissues, the wood frog's liver converts glycogen into glucose. Glucose acts as a kind of natural antifreeze that inhibits tissue damage in extreme cold. Breathing stops, heartbeat stops—a true state of suspended animation. The gray tree frog and the spring peeper, also found in Minnesota, have a similar capability.

The northern leopard frog has a different cold-weather strategy. It overwinters on lake bottoms. The animals gather atop one another in large, massed piles. Movement and metabolism are dramatically reduced. What little oxygen they require is absorbed through the skin directly from the water.

Leopard frogs sometimes breed along the fringes of deep lakes that are also suitable for overwintering. But because such waters typically are home to fishes and other predators dangerous to their eggs and especially to tadpoles, the frogs usually prefer shallower ponds like the one at the Ney farm as breeding sites. In the harsh Minnesota winter, shallow wetlands may freeze solid all the way to the bottom. But even short of that the ice can be so thick that the relatively small volume of water left unfrozen beneath it is soon depleted of oxygen. Leopard frogs therefore must make their way to deeper waters in the winter. Their migration can cover as much as a mile and the frogs generally retrace their route each spring and fall, traveling between the same breeding and overwintering sites year after year. How they find their way is not entirely certain, but they appear to rely mainly on scent and celestial navigation, with the summer sun guiding them home.

All of this played through Bob McKinnell's thoughts as he first took in the Ney pond. He immediately surmised that the pond was the frogs' breeding site. But it was too shallow for the animals to overwinter in it. That meant that for several months of the year the frogs would be found in a completely different corner of the local ecosystem. But where? There was no place anywhere on the undulating plateau of tilled fields immediately surrounding the pond that would provide an overwintering habitat. The only option—surprising but virtually certain in McKinnell's mind—had to be in the Minnesota River. On a straight line the river lies perhaps a half mile to the west of the Ney pond and at the base of a steep bluff—a cliff, actually—

that would seem an almost insurmountable obstacle to migrating frogs. Yet McKinnell and Helgen were eventually convinced that the river was the overwintering site—specifically a still, oblong backwater of the river known locally as Mud Lake, which sits right at the foot of the bluff.

All of this supposition was confirmed later that winter when a team of divers Reinitz recruited from a nearby college scuba club explored under the ice at both Mud Lake and the Ney pond itself. The divers found only one frog in the pond, presumably a lost individual that had somehow missed the migration. Down by the river it was another story. Mud Lake was shallow, too—so shallow that the divers had to take their tanks off their backs and carry them alongside as they swam through the murky, claustrophobic space between the ice and the bottom. This effort turned up nothing directly. But Cindy Reinitz and her students, watching impatiently from the bank, did. While the kids waited for the divers they scouted the shoreline and before long they had found frogs, quite a few of them, lolling beneath the rim of ice where seepages of groundwater running into the lake had left small pockets of open water.

So the story of the Henderson frogs went like this: In the spring the frogs left Mud Lake at ice-out, probably sometime in March, climbed the bluff, traversed the fields, and began their warm-weather sojourn at the Ney pond. There, frogs at every life stage—egg, tadpole, juvenile, subadult, and adult phases—would be exposed through direct absorption to anything that might be present in the water. As spring turned into summer the larger frogs would move out over the landscape to feed on insects, while the mostly vegetarian tadpoles remained in the pond eating algae. Either food supply—free-moving insects or algae—could conceivably be a source of accumulated toxins. In midsummer the tadpoles would begin to meta-morphose and join their predecessors in the fields, where they would become both predator and prey. Small mammals, birds, and snakes would reduce their numbers significantly. By late fall the frogs still alive—repre-senting only a small fraction of the thousands of eggs from which they came—would find their way back down the cliff to Mud Lake where, in a long, slow-motion, half-congealed state of hibernation, they rested through the winter as their bodies drank in God-knows-what from the blighted waters of the Minnesota River.

The idea that the deformities might result from river contamination was especially troubling to McKinnell and Helgen because it would imply an accumulation of some toxicant in adult frogs that was subsequently transmitted to their offspring in the yolk of the egg—a process biologists call "maternal transfer." That could mean the toxicant built up over time in the adults but was present in the environment in only very small quantities that could be extremely hard to detect.

Privately, McKinnell was unnerved by what he saw in the summers of 1995 and 1996. He told me later that it all reminded him uncomfortably of an earlier chapter in his life. In 1980, McKinnell had been asked to help evaluate possible chromosomal abnormalities among persons recently evacuated from the Love Canal neighborhood of Niagara Falls, New York. Love Canal sat on property that had been owned by the Hooker Chemical and Plastics Corporation during the 1940s and early 1950s. By the late 1970s it had become a "neighborhood of fear" in the words of New York's then-governor Hugh Carey. Residents fell ill as toxic residues leached from the ground and seeped through basement walls as a tarry sludge whenever it rained. In 1978, Love Canal was declared a federal disaster area when it was learned residents appeared to be experiencing increased rates of cancer, miscarriage, birth defects, and other ailments. The cause was believed to be the more than 20,000 tons of toxic waste—notably dioxin—dumped there by Hooker. In the spring of 1980, President Jimmy Carter declared a state of emergency at Love Canal and the federal government began a relocation of more than 2,000 residents at a cost of several million dollars. McKinnell was one of several scientists selected to confirm a preliminary finding that a significant number of residents in the neighborhood had suffered chromosomal damage.

Chromosomes, tiny, intricately folded strands of DNA and protein found in all living cells, are segmented into thousands upon thousands of genes. An organism's genes regulate development and many other vital biological functions. Minor chromosomal abnormalities—small breaks and other irregularities—occur with aging and are not at all unusual. They're caused by everyday exposures to a wide variety of natural and manmade compounds as well as sunlight. But more serious aberrations can result from exposure to acute toxins or potent carcinogens, and these may have serious consequences for the individual as well as for the normal development of offspring.

McKinnell had reported finding abnormal chromosomes in the Love Canal samples—a sensational and hotly disputed result at the time that was subsequently supported by other researchers. The experience remained vivid to McKinnell, who at one point feared his career was coming to a premature end because of the intense skepticism over his results. Thinking back on it after visiting the Ney farm, McKinnell decided to contact a former student, Deb Carlson, now a professor at Augustana College in Sioux Falls, South Dakota. He asked her to begin chromosomal analysis on frogs collected by the MPCA. "I thought maybe we were looking at an animal Love Canal," McKinnell told me.

At it turned out, Carlson never did see any unusual chromosomal problems in the Minnesota frogs. But after McKinnell mentioned it to me I looked up some old stories about Love Canal and the controversy surrounding the chromosomal analyses that had been performed on citizens there. I came across an amazing passage in a letter to the journal *Science* from Margery W. Shaw. Shaw was a geneticist at the University of Texas whose own chromosomal analysis of people at Love Canal suggested that, if anything, McKinnell and the others had been too conservative in declaring that they saw significant abnormalities. Those conclusions had early on been branded a false alarm in the pages of *Science*. But Shaw vigorously defended the findings, which she readily conceded were open to interpretation. The extent as well as the causes and possible effects of chromosomal abnormalities were always going to be to some extent judgment calls, she wrote. Shaw continued, arguing that people need to know that science cannot give them all the answers, even when issues may be grave:

> We should recognize our ignorance and uncertainties and try to help the regulators as well as the human subjects to appreciate the concept of probabilities rather than certainties. In our democratic society, perhaps we will decide that 500,000 deaths per year is an acceptable price for toxic chemicals in our environment, just as we have decided that 50,000 deaths per year is an acceptable price for automobile travel. On the other hand we may decide that 5,000 deaths per year is an unacceptable price for toxic chemicals. The scientists should provide the data and interpret the results; the public should decide.

The analogy is imperfect—many traffic deaths are caused not by automobiles per se, but rather by controllable factors such as fatigue, excessive speed, or alcohol consumption that usually involve some degree of personal choice. The pernicious effects of toxic chemicals are less likely to be the result of a personal decision. But Shaw's point seems beyond dispute. Risk is an inherent part of life. Science can sometimes describe the hazards we may encounter at our own hands, but it can never totally eliminate them. Like a lot of people in Minnesota at the time of the deformities outbreak, I assumed the responsibility for this sort of problem belonged to someone, some agency of government. Surely our environment was policed, and had been for decades—wasn't it? Once the government got involved the deformities would be investigated like a crime scene, wouldn't they? Scientists would identify the causes and direct a cleanup of whatever mess we'd gotten ourselves into.

Right?

The shock of what was happening in Minnesota was that it so violated the anticipation of rebirth that makes life bearable here, however tenuously. Minnesotans live entombed for half a year beneath the snows of a brittle, featureless winter—a wan season of faded light and depressed spirits that approaches an absolute zero of the soul. But no place on earth feels so alive as the Minnesota countryside in August. Spring and autumn are less bleak than winter but are riven by abrupt changes. Storms sweep unimpeded across the northern prairies, wet and cold and howling as one extreme is replaced by another. Summer, what there is of it, is a hot, transitory flash that jolts everything back to life. By August the transformation is total: The forests and grasslands sway densely in warm breezes beneath an indescribably blue sky. Row crops stand tall where only weeks before there was nothing but plowed earth. The young of the year abound. All creatures imaginable have produced a new generation in the same short window of time.

But in the quickening, blistered summer of 1995, a transformation of a different sort had taken place.

When the Fish Began to Walk

SOONER OR LATER, MOST TRUE STORIES IN BIOLOGY ARE STORIES about evolution. Evolution is the most important of what biologists call the "emergent properties" of the chemical constituents of the universe. But if you're not a biologist you're unlikely to think of evolution as a subject with any direct bearing on you. We all know that other kinds of organisms—dinosaurs, for example—have lived on the earth and disappeared from it. We learned in school that the human species evolved from earlier versions of upright primates and that we are closely related to other primates still living.

Biologists take a broader—and infinitely longer—view of evolution. The development of life reaches back in time further than we can readily comprehend. Earth itself is about 4.5 billion years old, a number so meaningless that the writer John McPhee had to coin a phrase to describe this time beyond imagining in his book on geology, *Basin and Range*. McPhee called it Deep Time, and the term stuck. We poor humans can fathom time on the order of thousands, or, if we stretch it, maybe even tens of thousands of years. Millions of years don't mean much to us; billions of years are utterly imponderable. Yet most of the life-forms that have ever existed and evolved on earth lived in Deep Time—further back than we can dream.

One way to try to comprehend the awesome abyss of Deep Time is to envision the whole history of Earth as taking place over the course of a single, twenty-four-hour day. On this scale nothing stirs until almost 6:00 A.M.—Earth is a dead place for nearly the first quarter of its existence. The morning and the afternoon and most of the evening are primitive but eventful: simple, single-celled bacteria evolve into more complex, multi-cellular organisms. Sexual reproduction brings still greater variety and diversification. Life partitions itself into the plant and animal kingdoms. Yet the day is all but gone before much of anything resembling modern life-forms develop. Dinosaurs don't show up until around 10:40 P.M.! Humans arrive on the scene after 11:58 P.M. Civilization, the whole great, passionate drama of recorded history, from the pharaohs to the Internet, takes place in the last second before midnight. Yet what happened in the long evolution of life through Deep Time bears directly on life as we know it now. Evolution defines the relationships among the species and even suggests some answers to the question: What might a deformed frog have to do with me?

———————————

Those first bacteria were like seeds. They appeared about 3.5 billion years ago, and from them grew the entire tree of life. The tree diverged and spread into countless different branches. Many of those branches ended or broke off in Deep Time—more life-forms have gone extinct than exist today. But those branches that remain intact in the present time constitute all the examples of life on Earth. Every living thing is thus related, at some point in time, to every other living thing, in exactly the same way that every branch and twig of an actual tree is ultimately connected to the trunk. Travel back far enough into Deep Time and you will find that each organism alive today has at some point in the past a common ancestor with every other organism, though they may be removed from one another by huge intermediate evolutionary steps. The more alike two organisms are, the more recently their shared ancestor lived. Conversely, the less alike two life-forms are the more remote in Deep Time is their common relative. The English paleontologist Michael Benton has noted, by way of example, that humans and chimpanzees are derived from a common ancestor that lived a mere five million years ago, but to find the common ancestor for humans and lettuce you would have to look backward some 800 million years. That's a very big difference, one you may readily grasp the next time you have a salad.

Evolution produces great changes from small increments occurring over long periods, yet despite the vastness of Deep Time, evolution is not always a slow, stately process. Nature sometimes double-clutches. There have been long stretches when there was very little new under the sun. But there have also been riotous upheavals—mass extinctions that all but wiped life from the face of the planet, as well as times when new species appeared at a breakneck pace. Evolution is uneven, prone to sudden hiccups. As a result, Deep Time is rife with fits and starts, and certain classes of organisms stand at dramatic crossroads.

Amphibians are one such group. About 350 million years ago a few bony-finned fishes began venturing out of the seas and onto land. Developing lungs to breathe the air was a relatively small trick. What was more important was how they learned to get around in their new environment. Fishes were the first vertebrates, and it turned out that having a backbone was as important in an evolutionary context back then as it is in a moral sense today. A backbone, plus its accompanying internal skull and skeleton, conferred a tremendous competitive advantage in the ancient forest—a dense, damp greenhouse filled with towering trees and giant ferns, and already patrolled by a wild array of invertebrates, mainly insects.

As the first land vertebrates, amphibians—which initially looked something like a salamander and were probably about forty inches long—started what would become the most advanced branch of the tree of life. Their skulls provided a protective enclosure that could accommodate the larger brains of species yet to come, while their backbone and skeleton made much larger animals feasible. Internal bones provided anchoring sites for limbs at the shoulder and pelvis. Though they remained partly aquatic, especially during initial development, amphibians started the really big show on land, launching the evolutionary lines for all of the terrestrial vertebrates, from dinosaurs to man.

But primitive amphibians were more than an interim step leading to the higher vertebrates. Amphibian evolution continued along on the branch of the tree leading to modern amphibians—a diverse group that includes frogs, toads, salamanders, and legless, wormlike creatures called caecilians that live mostly underground and which most of us have never seen. Modern amphibians possess a wide repertoire of natural histories. Most remain truly "amphibious," but some are totally aquatic and others are completely terrestrial. Some lay their eggs on land and skip the larval stage. A few even give birth directly to fully formed offspring.

In opening the way for the biggest and most complex organisms to evolve, amphibians brought many things to the party, but nothing that was more important than one central design feature: four limbs. The first land vertebrates were also the first tetrapods, laying down the basic architecture for every vertebrate since. Cows, giraffes, bears, iguanas, gazelles, brontosauruses, sheep, elephants, pigs, saber-toothed tigers, gorillas, frogs, and me and you—tetrapods every one. Even snakes had four legs at one time.

The tetrapod limb is a primary element of life as we know it. It provides support, locomotion, and, in the case of humans, an ability to hold and manipulate things that is as important to us as our highly evolved brains. This sentence wouldn't count for much if I could only think it. But I can type it and you can hold this book in your hands and read it. All this courtesy of the miraculous moment in evolution when the fish began to walk.

In turning fins into legs, primitive amphibians developed a sequence of genetically programmed biochemical processes that regulate the formation of the limbs. The tetrapod limb is among the most structurally complex organs in the body; its proper development requires an orderly, precise arrangement of tissues. A frog with a deformed leg or with the wrong number of legs is therefore not merely abnormal, it's an insult to 350 million years of evolution.

Deformed limbs are evidence that something has caused the undoing of an important component of the fundamental architecture of life—an architecture laid down in Deep Time and reiterated among the many species of vertebrates, including humans, ever since. Whatever that something is, it has done this work of unraveling at warp speed. In evolutionary time a hundred years—a thousand, even—is nothing. Instantaneous. Even allowing for the probability that some frogs in Minnesota were deformed before Cindy Reinitz and her students found them at the Ney farm, it all seemed to have taken place over the span of a few seasons at most. Just like that, one of the basic strands in the fabric of life, a thread extending back into the mists of time, had been broken.

Pffft.

Scientists are divided about the concept of "sentinel" species. But it's an appealing idea: Certain classes of organisms, because of their constant or heightened exposure, may be more sensitive to environmental stresses and therefore provide related species an early warning of environmental degradation. In theory, animals like frogs that commune more or less unprotected with their environment ought to be like the canaries that coal miners used to take with them into the earth to warn of poison gas buildups.

But skeptics find the canary-in-the-coal-mine analogy a poor one. Canaries detect the presence of poison gas before it affects humans not because of a delicate metabolism, but rather because, unlike people, canaries completely empty their lungs between breaths. The sudden presence of poison gas in the mine hits the canary at full concentration immediately, on the first whiff, whereas it is initially diluted by normal air retained in the lungs of the miners.

Of course, the miners didn't care why a canary reacted more quickly than they did to a potentially lethal gas—only that it did. A modern toxicologist would agree that that's useful information—but only because the potential toxicant was already known. In this example the canary, is a sentinel for a single environmental insult, one that is understood to affect both canaries and people. In the impassive argot of science, humans and canaries share a common "end point" when exposed to poison gases. The problem with a supposed sentinel species comes when you try to interpret a different negative response—or the lack of one—to an unknown factor. What would a miner think if his canary sang on brightly but its feathers fell out? Would that signal the presence of something harmful only to organisms with feathers? And a canary wouldn't be of much use say, in a factory where the workers were concerned about detecting the presence of a slow-acting carcinogen.

Toxicologists have learned there's no such thing as a sure thing when it comes to extrapolating toxic responses from one species to another. Some toxicants affect broad classes of organisms—all vertebrates, for instance—in more or less the same way. Others may produce similar effects in different organisms, but at varying doses. Still other substances may be highly toxic to one organism and have no observable effect on another seemingly closely related organism.

This was horrifically demonstrated in the early 1960s in an outbreak of human birth defects caused by the sedative thalidomide, which had been widely prescribed in Great Britain as a treatment for morning sickness in pregnant women. The drug was previously tested on pregnant mice and rats, where it showed no effect at all on their offspring. But more than

7,000 British infants with severe developmental abnormalities were born to women who'd taken thalidomide and it was hastily taken off the market. The birth defects affected a number of organs, including the intestines and heart. Some children were born without ears. But the most telling deformity was a condition called phocomelia, in which the long bones in the arms and legs were absent or severely reduced and the hands and feet grew out at the ends of these much-shortened limbs, producing flipperlike appendages. Judy Helgen once said that among people old enough to remember this episode it was common to hear them refer to Minnesota's deformed frogs as "thalidomide frogs."

The exact mechanism by which thalidomide caused such serious birth defects in humans remains unknown nearly forty years after the fact. But subsequent tests on monkeys did produce abnormalities similar to those in humans, and the absence of any effect by thalidomide on rodents is a striking example of how different species process the same materials differently. Which is what you should expect, say critics of the sentinel species concept, because when you talk about different species the key word is *different*.

The main difference between species is reproductive. By definition, organisms belong to different species if they cannot interbreed and produce offspring that can reproduce. Cocker spaniels and poodles are different breeds of the same species because they can mate successfully and their offspring—cockapoos—can also reproduce. A horse and a donkey, on the other hand, belong to different species. Even though they can interbreed, their offspring—mules—are sterile. Beyond this basic distinction of what constitutes a species lies infinite variation in the way the myriad organisms of the earth metabolize and respond to environmental factors. Indeed, such differences have become an important premise in the chemical mix of modern life, especially in agriculture. Farm pesticides, chemicals designed to kill bugs and weeds and fungi—but not crops or pets or people—rely on alternate responses in different species.

Differences among species are complemented by many similarities, however. Biologists call it *conservation*. Many species share genes—and the proteins those genes produce—and the effects of such shared genes are essentially the same in every organism. There is really one only genetic code. In loose terms, you can think of DNA as a kind of alphabet. Its letters can spell out the thousands of words that make up the encyclopedia of instructions needed to form an organism and keep it alive. Different organisms have different instructions, but may share many of the same words. So the same arrangement of letters in any organism will always "mean" the same thing regardless of what kind of living thing you are. Many kinds of metabolic processes and developmental sequences are thus conserved among species.

This was the chief worry about the deformed frogs. Limb development is believed to be a process that is highly conserved among vertebrates.

––––––––––––––––––––

Even so, say the skeptics, just because one kind of animal or even a whole category of animals seems to be negatively affected by something in the environment, it is no proof that any other animal is at risk. But, of course, the reverse must also then hold true: The apparent good health of any group of organisms is no assurance that all is well for other living things. Life seems to be full of similarities, but also full of exceptions. What can we make of it? When we look around at the world and observe its effects on other organisms we can infer...nothing? Can we ignore the canary lying on its back? What explanation would make us completely comfortable that a frog with six legs has no implications for us?

The answer seems to depend on how far you insist on extrapolating from a single observation. Take the dead canary or the deformed frog at face value, say the advocates of the sentinel species concept, as sure, inarguable indications of conditions that are inhospitable to canaries and frogs. Can these things occur without probable effect on anything else? What are the chances that an environmental stress would affect one, and only one type of organism? Why not accept a problem in one species as evidence for potential problems in other species—a true "early warning" requiring further investigation. Nothing more. Nothing less.

This only sounds obvious. Toxicologists have trouble with the sentinel species concept precisely because it only provides hints, not definitive answers, and toxicologists already live in a world of ambiguity. Toxicology is not an exact science. The question for a toxicologist is not what is poisonous and what is not. In sufficient concentrations, virtually all chemicals are toxic. What concerns the toxicologist is the concentration that really counts—the effective dose required to produce a biological response. The acute toxicity of a compound—that is, the amount that is lethal—is dose dependent. The more of the chemical an organism is exposed to, the more certain it is that it will die. Toxicologists call this relationship the "dose response curve." It will be different for every animal. And that's only the beginning of the uncertainties.

An elephant would likely tolerate a larger dose of a poison than a human. But not all toxic substances are acute poisons. Carcinogens are generally toxic as the result of chronic exposures that are additive over time, and their effects may not be detected for years or even decades. The nature and degree of toxicity of any substance can also depend heavily on the maturity of the organism exposed to it. Developing animals are in almost every way more sensitive to toxins than are adults.

The issue of chemical toxicity is by nature a fluid, sometimes subjective search for an answer that is never absolute. It's all relative. Everything depends on the organism involved, the toxic effect you're looking for, and the dose it takes to produce that response. You can get a sense of the difficulties inherent in any toxicological inquiry by considering the effects of something many of us would consider safe—even pleasant—to consume: a martini.

Provided that you're not an alcoholic, you can safely enjoy a martini or two every day without ill effects. Recent studies on alcohol consumption and heart disease even indicate certain potential health benefits may result from such a regimen. But, if you were to drink an entire bottle of gin in the space of half an hour it could kill you. Short of causing death, the toxic effects of drinking that much gin all at once would certainly be more than apparent the next morning.

Think about one additional factor that might enter the picture. Consider again the effects of having "just" one or two martinis each day, only now imagine that you are a pregnant woman. There is still no risk that you'll find this much alcohol toxic personally. But your developing fetus might. Even small exposures to alcohol during development can produce birth defects, including reduced head size and, in severe cases, significant mental retardation.

The variables of dose and sensitivity confront the toxicologist at every turn. Things are complicated even in the lab, where everything you're not investigating can be eliminated or controlled. In the field, the problems multiply dramatically. When Judy Helgen and Bob McKinnell tried to assess what might be happening to the frogs at the Ney farm they had no real idea where to start.

———————————

Despite what looked like a high-risk pas de deux between the frogs and two possibly contaminated bodies of water at the Ney farm—the pond and the Minnesota River—McKinnell and Helgen could only guess that the frogs there were being exposed to a toxicant in the environment. There were other possibilities, including disease or parasitism. It was even conceivable that they were witnessing some weird, previously unimagined effect of an incremental elevation in the level of ultraviolet radiation resulting from the thinning of the ozone layer in the upper atmosphere. Or maybe it was some combination of two or more of those factors.

Questions were raised about local manure spills, about an abandoned copper mine down by the river, and even about the proximity of high-voltage power lines and electric substations. An eccentric man in Washington State called Helgen on several occasions and went on and on about something he called "bioelectric magnetism" and a mysterious government project involving "the generator." He insisted that Minnesota was being "bombarded" and suggested Helgen check for local increases in admissions to mental hospitals. Helgen could only be glad the man didn't live close by.

Whatever the cause of the deformities, McKinnell considered the potential threat to humans very real. He believed in the sentinel species concept, and after decades of studying frogs he also believed that their complex life cycles gave them a higher exposure to environmental toxicants than many other species. At the same time, McKinnell cautioned everyone that dose, time of exposure, and the nature of an exposure were variables that would have to be taken into account in the search for any toxic substance that might be involved—and that it could be all too easy to jump to conclusions that would later prove off the mark. "Look, if you put frog eggs into human mothers' milk they will all die," he told me once. "Does that mean that mothers' milk is toxic to people? Certainly not."

But the link between frogs and humans was undeniable. Both were vertebrates. And that was enough to trouble the researchers from the very beginning. It was the same reason that McKinnell had long studied tumors and chromosomal changes in frogs. Many cancer-causing agents actually acquire their carcinogenic properties when they are processed in the liver, an organ whose main function is supposed to be detoxification. This perverse outcome, in which the liver "turns on" a carcinogen, is called metabolic activation. It works pretty much the same in frogs and in people. So does exposure to genotoxins, compounds that alter DNA. In the early 1980s, McKinnell and other researchers examined a spectrum of genotoxic chemicals for their effects on frogs. The evidence was clear that frogs react to many of the same DNA-targeting compounds as do humans. What was

particularly compelling, McKinnell told me, was that frogs sometimes produced abnormal offspring after exposures to extremely low doses of chemical toxins. So it might not take much of an environmental contaminant
for frogs to take an initial hit.

For a time, when all the sites appeared to be recently excavated, Mc-
Kinnell worried that a toxicant was literally everywhere underfoot. The
most likely contaminants that would reach a wetland at the surface—in
runoff or via aerial deposition—were pesticides that were formulated to
break down fairly rapidly in the environment. But a contaminant lying
underground had probably been there for some time. That would mean its
toxicity was long-lived, and the historical list of chemicals that fit that
description included some of the nastiest substances known: heavy metals,
PCBs, dioxin.

These concerns abated after Dave Hoppe confirmed the outbreak at
the Bocks' lake. The fact that he found abnormalities in multiple species
also appeared to eliminate the remote possiblity of genetic mutations, since
it was highly improbable that more than one species at more than one
location would go genetically haywire in unison. The discovery deformities
at CWB had thus been a kind of good news/bad news event. The good
news was that, unlike the Ney farm or Audre Kramer's pond, this was an
almost completely undisturbed, natural lake.

But that was also the bad news.

Rumors, Theories, Rules of Engagement

REETINGS. MY NAME IS JOE TIETGE."

The words appeared on Cindy Reinitz's computer screen when she checked her e-mail one day in Octoter of 1995, two months after her class made their discovery at the Ney farm. Tietge offered few formalities. He identified himself as a research biologist with the U.S. Environmental Protection Agency in Duluth and explained that he worked on a newly formed study team that was interested in fishes and amphibians. He hoped Reinitz could answer a few questions about the deformities she'd seen near Henderson. What species of frog was involved? What were the deformities like? Were any other agencies or universities working on the problem yet?

The importance of this last point, unstated in Tietge's brief message, was considerable. Tietge didn't know yet how significant the find at the Ney farm had been—or that it was only the first of an impending explosion of such reports that would come in from around Minnesota. But he was on the lookout for a problem to solve, and he needed one in which the unfolding story offered the EPA a lead role. It didn't hurt that Henderson was only a three-hour drive from Duluth.

Reinitz wrote back to Tietge describing the deformed frogs at the Ney

farm—she also began forwarding his messages to Judy Helgen. The outbreak seemed an incredibly fortuitous fit with the EPA's search for something to do on frogs. During the summer of 1995, the EPA's Mid-Continent
Ecology Division in Duluth underwent a reorganization. Tietge, who had
just rejoined the EPA following a two-year stint with an environmental
consulting firm in Colorado, had been assigned to something called the
Reproductive and Developmental Toxicology Team.

The EPA divides its mission into two broad categories. The health
groups within the agency are concerned with research and regulations that
protect human beings; the ecology side worries about other living things
and ecosystems in general. The Duluth lab is the agency's primary ecology
research facility. Built in 1967 as a freshwater biology lab, Duluth had
been incorporated into the EPA when that agency was formed in 1970,
and the place had literally grown up over the same period of time that the
science of environmental toxicology had come into its own. When scientists there began working on water-quality issues in the early 1970s, only a
handful of the myriad chemicals in commercial use had ever been tested
for toxicity and not many techniques existed for assessing the deleterious
effects of artificial compounds in aquatic systems. It had always been
assumed that aquatic organisms were not adversely affected by chemicals
entering surface waters and there were few limits on what industries and
municipalities could discharge into the environment. For twenty-five
years, researchers in Duluth and other scientists under contract to the lab
had developed screening assays and predictive models in the course of testing more than 10,000 chemicals, including unusually dangerous contaminants such as PCBs and dioxin. The lab had helped to establish some one
hundred water-quality standards that led to the regulation of what can
and cannot be released into aquatic environments. The enormous strides
made in cleaning up the nation's water systems in the last quarter century
were owed in large measure to the work done in Duluth.

But progress isn't always rewarded in a bureaucracy. By the mid-
1990s the Duluth lab was floundering, thanks to its own successes. Like
most federal research facilities, the lab had to rationalize its existence in
perpetuity. As water pollution eased it became less clear what the lab
should work on in the future. Every time an aquatic problem got fixed, a
new one had to be discovered to take its place. Restructuring was just
another rest stop on the lab's endless journey toward a new mission. To
maintain its status as a leading center for aquatic ecotoxicology the lab had
to continually reinvent itself, and in the process, reinvent the science it did.

One promising area of research was the emerging question of how chemicals influence development—the cascade of events by which a fertilized egg divides and differentiates and matures into an adult organism that is itself capable of reproduction. The biology of development is a profoundly complex chain reaction in which genes and hormones and mystifying biochemical communications among groups of cells build an organism according to the blueprint contained in its DNA. We take development for granted, but in truth we have only barely begun to understand how it works. Scientists know that some chemicals are bioactive and can interfere with developmental processes, but which chemicals do this and the ways they interact with the developing organism are still a sea of open questions. This is a fundamentally different kind of concern from the study of acute toxins and carcinogens that preoccupied the field of toxicology for years, and the science needed to test compounds for such effects is still in its infancy.

Tietge, who'd originally come to the Duluth lab back in 1986 fresh out of the University of Wyoming, had gotten back just in time to be picked for the team. A tall, soft-spoken Southern Californian with a penchant for jeans, string ties, and bluegrass music, he was thirty-nine years old. In his off hours, Tietge was building a house on an eighty-acre farmstead in the highlands above the North Shore about seventeen miles from Duluth, doing much of the work himself. As anyone who's ever done carpentry can tell you, it's 90 percent rote execution and 10 percent problem solving—and it's the problem solving that separates the good carpenter from the average one. Toxicology is similar to carpentry in those respects.

Tietge's working group at the lab was headed up by another researcher named Gary Ankley. Together, Tietge and Ankley had devised a strategic plan in which they would focus their research on reproduction issues in fish and developmental problems in amphibians. They had a lot of experience with fish. But amphibians were of particular interest because so little toxicological work had ever been done on them and because Ankley and Tietge figured there was a need to invent bioassays using amphibians. This was their strong suit. The development of new aquatic test organisms had been pioneered at the Duluth lab.

But they knew that they would need a "real world problem," as Ankley put it, to work on. Part of Tietge's job was to identify a project where lab experiments and a live situation in the field could be tied together. When they heard about the frogs at the Ney farm it seemed like an almost harmonic convergence of opportunity and mission. "We might have gotten involved anyway," Ankley told me much later. "But at the time it just happened that we had both a mandate and some resources to take the plunge." Tietge felt a little more certain that the deformities were something that they could not have ignored, no matter what. He told me they had been "surfing" for just such a situation, but the outbreak at Henderson seemed to be almost thrust in front of them. The swell rose quickly and the EPA caught the wave. It was easy. It was inevitable. And it was damned convenient. "I just thought we had to respond," Tietge said.

Whether the EPA had to respond or not, the fact that they did respond went a long way in defining the shape of the frog investigation. In science, as in most things, problems tend to take on the look of whatever it is you're out to solve. Tietge and Ankley didn't have to work very hard to see the deformities outbreak as evidence of a developmental disorder—just what they were looking for—and so that is what it became. It also seemed likely these disruptions of limb outgrowth were being promoted by some "xenobiotic" agent. A chemical. All of that seemed probable to everyone involved.

But in hindsight the enormity of the coincidence was always hard to miss. Frogs had been around for hundreds of millions of years. Yet they had gotten themselves messed up almost as if on cue for the EPA to swing into action. You couldn't fault Tietge and Ankley for seizing the moment. It was too perfect.

Tietge needed a place to start. Of all the hard-won principles learned at Duluth, none was more fervently adhered to than the requirement to work from informed, scientific inference. Inference, which can be construed as simply an educated guess about cause and effect, is the first step toward the formulation of a testable hypothesis—a testable hypothesis being the one essential ingredient in any endeavor that can legitimately be called scientific. Years of experience with aquatic toxicants had taught the researchers in Duluth the utter futility of trying to tease out harmful chemicals simply by analyzing raw water. In fact, toxicologists generally avoided working on so-called complex mixtures like natural water. They much preferred to test pure, unmixed compounds one at a time. A jug of water from CWB or anywhere else would contain hundreds, maybe thousands of compounds. Ankley and Tietge knew that collecting and analyzing water more often than not got you nowhere because it afforded no way to learn anything you didn't already know. Unless you easily found something in a complex mixture that you already knew was toxic, about all you were likely to find was more complexity.

This would have been news to almost anybody in Minnesota who'd given the deformed-frog outbreak a second's thought. Most people assume the chemical makeup of just about anything you can think of should be relatively easy to determine in this day and age. A suitable task for a machine. The chemical constituents of a sample of water, one would think, ought to be as transparent to a chemist as the water itself. But this is not true. Complex mixtures of substances in the environment—water or sediments or whatever—are stubbornly unyielding. They are full of stuff that is hard to separate from other stuff and harder yet to name. Complex mixtures can be taken apart, piece by piece—but only in stages. As soon as you bump into something that's an unknown it's easy to lose your way. And it's almost always the unknown you're after.

Tietge told me there were several problems in breaking down complex mixtures. He began with a familiar refrain. "The main thing is that you only find what you look for," he said. "Another is that you can only interpret effects that you understand to begin with. And even if you do find a chemical you're interested in, you may have toxicology data on it or you may not. Or the toxicology data that you have may or may not be relevent. You could spend a lot of time and energy and resources identifying hundreds of chemicals in a water sample and in the end have absolutely no idea if any of them were capable of causing developmental effects."

Instead of searching blindly for harmful substances in water samples, the EPA had come to rely instead on what they called an "effects-based" approach. This involved combining water analyses with initial toxicological testing on living tissues—whole animals like fathead minnows or sometimes special cell cultures that could be used to detect whether a chemical had biological effects. In simple terms, the EPA was accustomed to modeling toxic scenarios by trying to induce the effect seen in the field on a similar species in the lab. If water from the field caused the same kind of problem under controlled conditions, then they could begin working on what was in the water that produced the effect.

Tietge recalled several years after the fact that they had briefly considered a kind of brute force frontal attack on the deformities problem. He and Ankley debated whether to send a tanker truck out to one of the affected wetlands to collect thousands of gallons of water to bring back to

Duluth. "Then we were going to throw in some frog eggs and see what happened," Tietge said. They didn't do that, although Tietge admitted to me that he never stopped thinking it was a good idea.

Not surprisingly, Tietge doubted that the MPCA or anybody else would get very far simply by looking for connections among the various locations that were producing deformities. The idea that some common factor would turn up in all the affected wetlands wasn't an inference—it was a blind hope. Instead, what seemed to be called for was an understanding of how abnormal limbs grow and what environmental interference with those mechanisms could induce the deformities. By the summer of 1996—after months of looking into what was known about frog deformities and, more important, what was not known—Tietge decided to organize a meeting. He planned a two-day workshop in September that would lay out the data on frogs, deformities, and the science of limb development—a conference he envisioned would conclude with some sort of game plan for the EPA's own initiative.

This was a proven strategy—the EPA routinely consulted with other scientists, especially academics, as a way of developing leads on complicated problems. Tietge knew that meetings were often a shortcut to great stores of information that were buried in the literature, and that the brainstorming that took place when you got a lot of scientists together often led to jointly agreed-to inferences that could then be investigated. He knew, too, that real payoff would be access to a lot of information that was not yet in the literature—speculation, data from work in progress, and so on— that scientists would talk about freely long before their findings had been finalized and subjected to peer review.

I didn't know Tietge, but I heard about the meeting a few weeks before it happened and phoned him. I explained that I was working on a story about the deformities outbreak. He seemed a little taken aback but was polite and tried to be helpful. He named some of the people he expected to attend, including Judy Helgen, Martin Ouellet, Dave Hoppe, and a biologist from upstate New York who was the author of the 1990 paper that purported to explain frog deformities as a natural occurrence caused by parasites. Bob McKinnell had been asked to come, too, but had oddly declined. Tietge sounded annoyed by this. In all, Tietge expected nearly sixty scientists from around the country to attend. When I asked Tietge if I could sit in on the workshop he didn't answer for a second. Then he chuckled and said, "Well, I guess I can't stop you."

The first North American Workshop on Amphibian Deformities commenced at 8:00 A.M. on September 25, 1996, in the warm, brightly lit Great Lakes Ballroom in the basement of the Holiday Inn in downtown Duluth. Participants circled the breakfast buffet in an expectant hush, a steady clinking of china echoing in the hallway as people siphoned coffee into their cups from tall, silver-plated urns. A few of the scientists murmured greetings to one another as they helped themselves to pastries and pinned on name badges before taking their seats. Many of the academics chewing happily on the EPA's doughnuts had mixed feelings about the meeting. Privately, a lot of them thought of the agency as an arrogant, meddlesome bureaucracy run by scientists whose credentials were less impressive than their own. In the months to come I would often hear EPA researchers referred to derisively as "the Feds" and would listen as academics bristled at what they saw as the EPA's presumption of authority.

The conference room proper was arranged with a large screen at the front. A lectern stood off to the side and a slide projector sat perched on a stand in the middle of several rows of tables—the standard scientific meeting layout. It was pleasantly calm inside the Holiday Inn. Outside, the weather was blustery. A powerful gray swell was running in off Lake Superior ahead of an icy wind. The waves pounded the waterfront and sped down the ship channel in steep ranks beneath Duluth's famous aerial lift bridge a few blocks away.

What transpired over the next two days was, for the most part, a little less exciting than the tingly Duluth weather. Much of it was certainly harder to understand. Tietge called the meeting to order in a soft voice. He smiled self-consciously at the audience and seemed somehow a little bemused, as if the whole idea of such an ambitious gathering was out of proportion to the issue at hand. I'd had this feeling, too, on the drive up to Duluth the day before. I couldn't decide which was more surprising—the outbreak of deformities popping up around Minnesota or the EPA's sudden interest in it. Tietge had a plush toy frog with him and announced his intention to use it as a stage hook by waving it in the air at anyone who went on too long.

Tietge said the EPA felt it had an obligatory stake in the deformities outbreak, even if it was as yet impossible to say exactly what was happening or how serious it might be. In fact, he went on, those very uncertainties elevated the agency's concern. He said that since the deformities had been discovered over a large area of Minnesota and were seen occurring in a variety of different land-use settings, the outbreak had become progressively more baffling. With the reports increasing, the EPA felt compelled to get involved. The suggestion implicit in this was that the EPA was taking charge of the investigation, a suggestion that didn't seem all that necessary under the circumstances. Nor was it likely to endear him to anyone present. Tietge acknowledged that his agency's level of concern over the deformities was not universally shared in the larger scientific community, and noted that some scientists were dismissive of the problem, if it even was a problem. Others, he said, saw it as one more indication that it was almost too late for humanity to save itself from environmental disaster. "I believe there are some people out there who think *they're* growing extra legs," he said.

Dave Hoppe, Martin Ouellet, and Judy Helgen stepped to the lectern in turns to describe what had been seen in the field, with the rhythmic whir and *ka-chunk* of the slide projector punctuating their respective reports. There didn't seem to be any question that something unusual was happening across Minnesota and southern Canada. But the real business of the conference was to propose causes and the means to investigate them. Presently, two were put on the table: chemicals and parasites.

Think of science as a great city, with biology one of its boroughs. Along broad boulevards stand skyscrapers of theory, the pillars of our understanding of the nature of life. Walk down a side street into one of the neighborhoods and you will find that everything is studied by somebody. All that is alive or that has ever lived is a subject of inquiry, from bacteria to blue whales to giant redwoods. Go a little farther, into the cul-de-sacs and back alleys and you will discover that living organisms are further divided, down to the cellular level, and then deeper and deeper still, with each particularity accounted for, every subunit of the machinery of life parsed and reduced to a swirling collision of molecules. This is how it is in the corner of the city where they work on legs.

For more than a year, the investigators in Minnesota had searched in vain for a common denominator among the sites where deformed frogs had been found—some agent in the water or on the land that was obviously out of kilter. Joe Tietge and Gary Ankley had a different take on what to do. What interested them was the only thing the sites clearly did share with one another: deformed frogs. In the symptom, they reasoned, lay the most important clues as to the cause. A natural experiment of sorts was already complete. Abnormal limbs were the data. Any investigation into what caused them would in the end have to reckon with the mechanics of limb development. Why not start there? Tietge had sought out an expert on limb development to come to the meeting and he'd found one of the best—Ken Muneoka, chairman of the biology department at Tulane University.

The limbs are the last major organ system that develops in a vertebrate animal. By the time they begin to form, the embryo has already solved the biggest and most mysterious problems in development: It has determined its right from its left. Also its head from its tail and its front from its back. Every higher organism starts life as a single, spherical, fertilized cell. Infinitesimal asymmetries in its makeup guide what happens from the moment life begins. The one cell cleaves into a bundle of smaller cells that then begin to multiply and differentiate according to subtle but powerful biochemical signals. Within this organizing scheme, millions of cells eventually locate within specific tissues and assume specialized functions. We take this three-dimensional miracle for granted, but it is a miracle all the same. In 1986, Muneoka and his mentor, Susan Bryant of the University of California Irvine, marveled at the speed and surety of development in an article they wrote for the journal *Trends in Genetics*:

> Many years of evolution underlie the structure and organization of all living things, yet in each generation this unique arrangement of differentiated cell types of an organism is created afresh in a matter of days or weeks, using the genetic information present in the fertilized egg.

Limb development recapitulates the kinds of cell differentiation and orientation that have earlier taken place in the body of the embryo. That is, a leg undergoes another sequence of pattern formation and growth that produces a fully formed, three-dimensional structure.

Pretend it is New Year's Day. Imagine you are attending the Rose Bowl in Pasadena to cheer on your alma mater, the University of Minnesota Golden Gophers football team. This is imaginary. As a member of the Gophers' rooting section, you and several thousand of your old classmates have been assigned to occupy one side of the stadium and to participate in the halftime show by holding up colored placards. As you stand around outside the turnstiles before the game you are functionally indistinguishable from one another. But as you enter and are given your placard and directions to your assigned seat differences and patterns are being established. Most of the placards are maroon or gold—school colors. A few are white. Everybody gets just one placard, and to you, your one piece of the puzzle doesn't mean anything.

Collectively, though, a complex structure will emerge if the placards are correctly distributed and arranged. When the signal comes at halftime and everyone holds up their placards, one entire side of the Rose Bowl will be tranformed into an improbable work of art: an immense depiction of Goldie Gopher, the team's trusty mascot, frozen in midstride, carrying a football and decked out in maroon-and-gold regalia. Beneath Goldie, in elegant white letters, will be the words "Go Gophs!" You won't see any of this, just your one little colored placard and maybe a few others nearby, at least not until you get home and watch a replay of the tape shot from the Goodyear blimp. But to the people on the other side of the Rose Bowl and those watching on TV, the individual placards will all merge into a complete, perfect image.

This is a reasonable approximation of the way limbs develop. In the darkened basement of the Holiday Inn, Ken Muneoka carefully explained how the limb assumes its proper shape and form. The process begins, he said, when cells in the embryo begin to bunch up at sites where the limbs will grow. Within these assemblages of cells resides the biochemical machinery needed to make a complex, multitissued organ.

Under normal conditions, limbs can only grow out from the flank of the embryo, at points on opposite sides of the body where they will connect to the shoulders and pelvis. Each leg (or arm or wing) begins as a limb bud located just underneath a ridge of tissue that forms on the surface of the trunk. The limb bud is packed with undifferentiated "general purpose" cells that are like those Gopher fans waiting at the turnstiles.

In amphibians the outer ridge of the limb bud is less well defined than in other vertebrates, but this bulge of "permissive" epidermis begins an intricate chemical conversation among the cells in the limb bud—a signaling and cross-signaling sequence that does the same thing those halftime planners handing out the placards to the fans as they file into the Rose Bowl do. Like the fans, who acquire an identity in the eventual pattern by being designated as one color or another, the cells differentiate into different types in order to form various tissues. Muscle. Bone. Skin. And so on. The cells migrate and divide and organize themselves. As the limb bud grows out from the body, the cells inside it continue to multiply and move to their correct positions—in the same way everybody at the big game takes their assigned seats.

The limb pattern reaches completion while the limb is yet quite small. A final stage of programmed cell death occurs in the paddle-like structure at the end of the limb and the toes (or fingers) are separated from one another. After that, everything just gets bigger as the animal grows. Showtime!

Scientists have been working for decades to understand the processes that regulate differentiation and pattern formation in the vertebrate limb. Some of the early experiments were worthy of Dr. Frankenstein. It was learned, for example, that if the ridge of epidermis at the tip of the limb bud is removed, the limb will cease developing at whatever stage this removal occurs. The only exception to this among vertebrate organisms is the salamander, which can regenerate all or part of a limb at any stage of life. Frogs also have a regenerative capability in the early larval stage, but it declines rapidly and disappears as the onset of metamorphosis approaches.

Grafting experiments on salamanders led to important understandings of how limbs develop, albeit in results that often seemed utterly confounding. For example, when a limb bud is amputated from one side of a salamander and then grafted onto the stump of a similar amputation on the opposite side—so that the "back" and "front" are reversed 180 degrees—the swap produces a surprising result. Instead of the limb simply growing out backward as you might expect, three limbs grow out instead, in a sort of T formation in which the two arms of the T are mirror images of each other.

This effect, achieved in experiments derived from earlier work on cockroach limbs, demonstrated that differentiated cells "know" which kinds of cells they are supposed to adjoin. When they are confronted with cells from a different position in the limb, the opposing cells go to work and "infer" all the missing positions that should otherwise separate them. Thus when the leg is swapped from side to side and the front and back edges are reversed, cells in the transplant and cells in the limb stump to which it's attached invent the regions of limb that seem to be missing and two extras are produced. In the mid-1970s, Susan Bryant was one of several collaborators who developed a theory called *intercalation* to account for this effect.

Intercalation is an important if hard-to-grasp property of the developing limb. It demonstrates the ability of cells to communicate with one another and to share information about where they are in the limb and what kind of tissue they're supposed to be. In our Rose Bowl analogy, you could achieve something like intercalation by giving everyone a copy of the master plan for the halftime show. Then you'd know what color placard the person sitting next to you is supposed to hold up. If the Gophers fell behind and someone drank too much beer and had to leave right before halftime, the people around that seat could call somebody else in to hold up the placard and the show would go on. Intercalation at work.

If some people accidently went to the wrong seats the area around them would look incorrect or incomplete based on the pattern they were supposed to be part of. If they were pushy enough to shift people around according to their understanding of the plan, then elements of the final image would turn up in the wrong place. Goldie's leg might end up sprouting from his forehead. Intercalation again.

Muneoka explained that more recent experiments have begun to identify the specific genes and molecules that determine the three axes—base to tip, front to back, and top to bottom—that must be defined in the limb as it forms. Muneoka said that one of the most important of these biochemical factors is retinoic acid, a hormone derived from vitamin A, which is critical in initiating limb bud development. Retinoic acid also plays a role in regulating later outgrowth of the limb. Muneoka emphasized that the development of abnormal limbs—especially extra limbs—most likely involved a disturbance of the limb field at a very early stage. By way of example, he said that early-stage exposures to excess amounts of retinoic acid had been shown to cause the formation of extra or "supernumerary" limbs in both mouse and chicken embryos. Other experiments had demonstrated that suppression of retinoic acid synthesis in developing embryos could cause limbs not to form at all.

One of the most startling effects of retinoic acid was discovered in an experiment done on frogs, first in India in 1992, and then a year later when it was repeated by a British researcher at King's College at Cambridge. Researchers amputated portions of the tails from tadpoles that were then put in a solution containing retinol palmitate, a retinoic acid derivative. The larvae would normally attempt to regenerate their tails.

But this time, instead of regrowing tails, the tadpoles sprouted clusters of extra legs from their tail stumps. This bizarre effect became known in biology circles as the famous "tails-into-legs" result.

Muneoka urged the deformities researchers to concentrate their efforts on searching for causes that could disrupt the basic mechanisms of limb formation. The surgical rearrangement of cells in developing limbs showed that abnormalities could result from a mechanical disturbance in the limb field. But Muneoka thought it more likely in the case of the deformities being reported from the field that some chemical perturbation was involved, very possibly some "retinoid" that was an analog or mimic of retinoic acid. A retinoid—or more than one—would make a good prime suspect in the deformities outbreak for a number of reasons.

Retinoic acid, like other hormones, can "turn on" specific genes by binding to a receptor in the cell. There are two distinct receptors for retinoic acid, and both are *nuclear receptors.* Unlike other receptors located at the cell's surface, nuclear receptors lie deep inside the cell, adjacent to the nucleus. They are extremely sensitive—biological hair-triggers. When a nuclear receptor is activated by its hormone, or something that looks like its hormone, it switches on a direct genetic response within the DNA segment it targets. It doesn't take much retinoic acid to elicit a response. The hormone functions normally at extremely low concentrations. A little too much, or a little too little, will throw a developing limb off track.

Retinoic acid also fits neatly into an emerging research preoccupation in the field of toxicology: endocrine disruption. Endocrine disruption had lately gained a lot of currency, thanks to an influential 1996 book called *Our Stolen Future*, which proposed that hormones and hormone-mimics in the environment are wreaking developmental havoc in wildlife populations and in humans. Endocrine disruption was already being implicated in increasing rates of cancer and birth defects, as well as in decreasing sperm counts around the world. The list of potential disruptors—hormone mimics—was a veritable Who's Who of environmental contaminants: industrial solvents, PCBs, dioxin, many pesticides, and even some kinds of plastics. Finally, and this was very important, there was an abundance of experimental results that showed how retinoic acid could cause abnormal limbs to form. It was a known bad actor.

Here was a concrete lead—albeit one wrapped up in a complex overview of limb development so dense it left just about everyone in the room at the Holiday Inn shaking their heads. If the concept of limb induction and cell-to-cell intercalation seems unfathomable to you, take comfort in the fact that it was to many of the scientists gathered in Duluth, too. When Muneoka finished and the meeting adjourned for the morning, I went to lunch by myself to collect my thoughts, which were jumbled. On the way back I ran into Dave Hoppe and asked him about Muneoka's presentation. He nodded his head. It was very important, he thought. Somebody needed to get to work on the retinoic acid angle. But it wasn't going to be him. "Me?" he said. "Hell, I wouldn't know a retinoid from a hemorrhoid."

———————————

As it happened, most of the speakers at the meeting had ended up sitting close to one another, on the left side of the room. Dave Hoppe was just behind Martin Ouellet who was right in back of Judy Helgen. When everyone got back from lunch I noticed that Ken Muneoka was once again seated next to a tall, professorial-looking man with a salt-and-pepper beard. He wore a tweed sport jacket with patches at the elbows and his glasses hung from one of those cords that go around your neck.

It was Stan Sessions, currently a biology professor at Hartwick College in Oneonta, New York. He and Muneoka knew each other well. They'd done postdoctoral research together in Susan Bryant's limb lab at Irvine in the 1980s. It was during that time that Sessions had linked frogs with abnormal legs to infections by a common parasite. Sessions and his collaborator, Stephen Ruth, had published their findings in the *Journal of Experimental Zoology* in 1990. "Sessions and Ruth," as the paper was referred to in the usual scientific shorthand, was the best-known publication on the subject of frog deformities in existence.

Sessions had a long-standing interest in both amphibians and limb development. Prior to starting his postdoc at Irvine, Sessions had received his Ph.D. at Berkeley, where he studied under Dave Wake, a member of the National Academy of Science who is arguably the foremost herpetologist in the country and certainly the best known. Sessions worked mainly on tropical salamanders; the deformities outbreak in Minnesota had unexpectedly revived a subject he thought he'd closed the door on years ago.

As Joe Tietge called the meeting back to order, Sessions loaded his slides into the projector, gathered up his notes, and headed for the lectern. The lights went down again and he got right to the point.

Limb deformities in frogs were not a new phenomenon, he said, emphasizing a point Dave Hoppe had made earlier in the day. There were numerous reports of outbreaks of frog deformities in the literature. His own encounter with the phenomenon occurred in 1986, when Stephen Ruth called him in to help evaluate what was happening in several ponds in Northern California near the town of Aptos. Large numbers of frogs and salamanders with extra legs had been discovered there during a routine assessment of the salamander population, which was endangered. Ruth had previously observed multilegged salamanders at a pond in the same area twelve years earlier. This time he wanted to know what was causing the deformities. Sessions had obliged, and after a careful investigation of samples collected in the field, he improvised an imprecise but imaginative experiment. He believed that he then understood what was going on. "In this case," he said, "we're pretty sure we found the cause."

Sessions showed slides of some of the historical examples of frog deformities and then moved directly into a series of photographs of the amphibians he'd examined. The frogs had been prepared using a common technique for evaluating skeletal structures called clearing and staining. The procedure involves skinning and evisceration of the frog, which is then stained to highlight the skeletal elements. Bone turns bright blue, cartilage a deep red. Finally, the animal is chemically treated to make the remaining tissues transparent. The result is a kind of glassy "visible frog" with a brilliantly hued skeleton embedded within a ghostly body outline.

Sessions's cleared-and-stained specimens were riveting, and they had a very different effect on the audience than the total gross-out of Dave Hoppe's pictures of living frogs. A chaotic, writhing mass of slimy legs in a live frog looked like—a mess. But when a similar frog was cleared and stained, its fleshy dysmorphologies disappeared and all you saw was an intricate, almost elegant stick-frog—a lovely colored boneworks. They looked beautiful, if maybe a little off.

In Sessions's study he examined 280 tadpoles and newly metamorphosed froglets, plus nearly 6,000 salamanders from varying age classes. The frogs were *Hyla regilla*, the Pacific tree frog, a small greenish-brown animal that despite its name is not notably arboreal. More than 70 percent of the frogs Sessions looked at exhibited abnormal limbs, mainly extra rear legs. One frog had a total of twelve hind legs. The salamanders were less severely affected, but almost 40 percent of the younger specimens showed limb defects, primarily extra digits and feet but also shortened hind limbs. Both the frogs and the salamanders exhibited a wide range of other deformities, including essentially all the types of abnormalities that had later been found in Minnesota and Quebec, although in quite different ratios.

At the time of his original investigation, Sessions had not considered extra limbs and missing or abbreviated limbs of equal interest. In a footnote to the field data in "Sessions and Ruth" he all but dismissed missing limb structures in the salamanders as irrelevant, writing dryly that they "could easily be explained by attempted predation." But the paper made no mention of what the likely predators were or whether they'd been found at the ponds or why predation didn't also explain the missing limbs seen in the frogs.

In several of the photographs Sessions put up numerous small, blackish dots were gathered in dark clouds in the hind limb region of the frogs' torsos, seemingly clustered in around the pelvis. Sessions waved circles around them with a red laser pointer. "You may be noticing these," he said teasingly. "I see them, too. I'll come back to them in a minute."

Sessions let this hang briefly before continuing. One of the things that I heard often about Sessions—that day and on many subsequent occasions—was that he was a gifted speaker. This was not necessarily an unqualified compliment and people who offered this opinion usually conceded they were envious of Sessions's persuasive abilities but felt that he advanced his argument more by force of personality than by credible evidence.

Certainly, his presentation that day in Duluth seemed smooth and uncommonly assured. Sessions exuded certainty on seemingly every point, and his talk brimmed with confident exaggerations and blunt assertions. He struck me as a man who quite plainly believed that what he was saying was not only true but in fact rather obvious—so obvious that anyone who disagreed with him was either stubborn or slow on the uptake. There was nothing subtle about the snub-nosed phrases that went off in his talk like little explosives. At several points, for example, he introduced new concepts by saying, "As we know..." and judging by the rapt silence in the room it seemed pretty clear that not everyone present did know what Sessions expected them to.

Sessions's presentation flowed inexorably toward a conclusion that was at odds with the prevailing opinion among other members of the audience, which was that an environmental catastrophe might be at hand. No, Sessions warned, what had been seen happening to frogs in Minnesota actually looked more like business as usual in nature. Those little black dots were the key. Those, he now explained, were encysted parasites, a kind of flatworm. The identification of parasites is a tricky business, he said, even among trained parasitologists, which he was not. But Sessions had tentatively determined that these appeared to be a digenetic trematode, *Manodistomum syntomentera*. The life history of a digenetic trematode, he said, is very interesting.

Sessions clicked the frame advance on the projector and a picture of a coiled garter snake loomed on the screen.

Garter snakes, he said, were the primary hosts for *Manodistomum*, which like other digenetic trematodes live and breed in the gut of vertebrate carnivores. But this parasite also has a great many discrete developmental stages and passes some parts of its life cycle in one or two other "intermediate hosts," waystations on a circular journey that begins in a snake and eventually returns to a snake, where it starts anew.

Sessions showed a slide of *Manodistomum*. It looked flattened, two-dimensional, and more or less egg-shaped. The picture had that grainy, microscopy look I remembered from high school biology. It made Manodistomum look like any of those sort of roundy, flat, amoebalike things you can't see with the naked eye but which your teacher assured you were floating around by the zillions down at the old swimming hole. Sessions pointed to two adhesive organs at the oral and ventral openings that he called "suckers."

When the trematode reproduces, Sessions explained, its eggs pass from the garter snake in feces, at least some of which end up in pond water. There the eggs hatch into an initial larval stage that infects aquatic snails. Inside the snails, the parasite continues on through several more developmental phases before bursting out—in a process called "shedding"—and reentering the water, this time as free-swimming, more advanced larvae called cercariae.

As I was trying to form a mental image of a submarine wave of itty-bitty teenage flatworms spreading out through the tea-colored waters of your average, harmless-looking Minnesota wetland, Sessions said that what happened next was that the little buggers made a beeline for the nearest tadpole or immature salamander.

"And, like any self-respecting parasite," he added, "they head right for the alimentary canal, preferably up the cloaca." But the trematodes didn't necessarily need an orifice to get inside an amphibian; they were quite capable of boring right through the skin. Apparently, they did both.

Once inside an amphibian, the cercariae took up residence. They might migrate around a little for a while, but soon enough they settled down somewhere in the soft tissues, curled up into a small wormy ball, and became fixed in position and encysted. This dormant, resting state is the "metacercarial" stage of the trematode's life cycle.

Sessions said that in their examinations of frogs from the ponds near Aptos metacercarial cysts had been found throughout the bodies of the infected amphibians, but that they tended to be concentrated in the pelvic region near the base of the hind limbs. He described these findings in graphic, war-zone terms. The cercariae ripped into pelvic tissues, including the developing limb bud, crowding together and *jamming up* the field of cells dedicated to limb outgrowth. The infestation, he said dramatically, was massive and did extensive *damage*. All this looked very suspicious, to say the least. And because no obvious chemical contamination had been discovered in tests on the pond water near Aptos, Sessions had wondered if the cysts weren't somehow involved in causing the formation of extra limbs.

At first, Sessions thought the encysted parasites might be secreting something—maybe even retinoic acid. But the more Sessions studied the extra limbs the more they reminded him of something else. Many of the "supernumerary" limbs were mirror images of the primary limbs they had grown alongside. This was the same sort of mirror-imaging symmetry seen in the well-known contralateral grafting experiments in which limbs were transplanted from one side of a salamander to the other with their front and hind edges reversed. Sessions showed a slide of the crazy, three-piece salamander leg from the graft.

"To be accepted into Sue Bryant's lab you have to be able to do this experiment," Sessions said, "so that you can see with your own eyes that it works."

Sessions wondered if something similar weren't happening to the frogs and salamanders at Aptos, some sort of perverse natural disturbance of the limb bud tissues in developing amphibian larvae that produced supernumerary limbs. Suppose the metacercarial cysts didn't secrete anything at all and instead simply rearranged cells in the emerging limbs in such a way that their developmental blueprint got scrambled. Could this cause limb cells to lose their place and, via intercalation, form extra legs? Sessions decided on an experiment to find out.

In the lab, Sessions implanted tiny resin beads into the limb buds of larval frogs and salamanders. The beads were about the same size as the metacercarial cysts. Sessions inserted a few of the beads into each animal—he could only get as many as five or so into any single limb bud—using fine, very sharp surgical forceps and simply forcing them down into the tissues. This rough treatment, he supposed, would be the equivalent of the rending of flesh that happened when parasites bored into the amphibians in the ponds. As a check, he performed the same forceps procedure on a set of control amphibians without inserting any beads.

The results of this experiment had been hotly debated ever since they were published in "Sessions and Ruth." But in Sessions's mind the data were conclusive: Nine out of forty-four animals—20 percent—developed "substantial limb abnormalities including duplicate limb structures" following the implantation of beads, while only a single small defect showed up in a digit of one of the control animals.

What the experiement showed, Sessions said, was that tissue damage from parasite attacks on developing amphibian limbs initiated a regenerative response that was confused by the presence of a solid object in the limb bud—the metacercarial cyst. This physical displacement of cells caused positional confrontations and, in turn, an intercalation in which whole new limbs were "inferred" by groups of cells that determined something was missing and proceeded to fill in missing pattern.

The actual data that appeared in "Sessions and Ruth" could be seen as more suggestive than conclusive. The phrase "substantial limb abnormalities" wasn't especially precise, while the photographs of the deformities elicited by the beads showed only hand and foot duplications—nothing that approached a whole extra limb. Even so, Sessions believed the case was made, and writing it up in his paper he used characteristically emphatic terms. The situation at Aptos had undoubtedly arisen as a result of a "sporadic and localized population explosion of trematodes and/or one of its hosts." With a nod to the many historical reports of multilegged frogs, Sessions wrote "our hypothesis is sufficient to explain the naturally occurring extra limbs" that had long been observed in frog populations. He suggested that anyone investigating a report of multilegged amphibians needed to explore the involvement of parasites.

———

One of the questions many people had raised about all of this was why. Why would a parasite cripple its host? A few weeks earlier I had put this very question to Bob McKinnell—who was well aware of "Sessions and Ruth." He was skeptical all the same.

"I'm not an expert on parasites," McKinnell said. "They're a possibility and some people think they are responsible for these kinds of deformities. But parasites evolve and adapt to the life cycle of their host. Normally you would not expect a parasite to kill its host. Well, the abnormalities like we're seeing are not friendly to the host. They're deadly. So my question is this—if frogs have lived forever with parasites, why would this relationship fall apart in a very short time? It just seems illogical to me. I'm not saying that it's not parasites. But if it is, I think it must be parasites plus something else."

This was the point that seemed to excite Sessions the most. It was reasonable to ask why a parasite would disable its host. But there was, in his view, a perfectly reasonable and actually quite elegant answer. In the harsh world of *Manodistomum* a frog is a dead end. To stay alive and reproduce and send another generation of baby parasites off into the future, *Manodistomum* has to get back into the belly of a snake. What better way to get there than onboard a frog that gets eaten by a snake?

It was a fact, said Sessions, that the multiple legs and other limb abnormalities in the amphibians at Aptos were a serious handicap that reduced the animals' locomotive capabilities and made them much more vulnerable to predators—including garter snakes. Over time, nature would have expressed an ineluctable preference for strains of *Manodistomum* that did the best job of inducing extra legs, compromising their intermediate hosts in a way that made them a more probable meal for their primary hosts. Parasites, snakes, snails, and frogs—all had come together in what Sessions called a coevolved system.

One little detail had to be explained. All of the frogs at Aptos, including the normal ones, were loaded with parasites. Why didn't *Manodistomum* cause limb abnormalities in every amphibian it got into? This was easily answered, said Sessions, though in the light of later developments the answer was maybe a little too easy. Parasite cysts could only cause leg defects if they were present during the short, critical window of time when the limbs were developing. Parasites that infected a later stage tadpole or a young frog would have no effect on the limbs.

As Sessions finished up his talk I thought I could feel the audience slumping in their seats. Was there really going to be any need for a second day of deliberations? I was making notes and silently composing a lead for my *Washington Post* story that would tell the world parasites were causing frog deformities in Minnesota when Sessions said something out of the blue that stopped me.

Having made the case for parasites for the better part of an hour, Sessions abruptly tossed it aside just before going back to his seat. Picking his words with difficulty, Sessions alluded to "a problem" he saw with the Minnesota data.

Not all of the deformed frogs in Minnesota had multiple limbs, even though the ones that did had generated a lot of excitement. Parasites were "always there," he said ruminatively, "more or less in the background." But there might be something else going on, too. "I came here convinced that parasites were the cause of deformed frogs. But what I've heard today makes me think that environmental degradation is somehow contributing," he said finally.

This wasn't quite the same as saying that chemicals might be the cause of the deformities in Minnesota, but it came dizzingly close. I felt as though I'd just watched someone with a winning lottery ticket rip it up in front of a crowd.

Sessions seemed to doubt that the data from Minnesota matched his explanation for frog deformities, but he might just as well have made the opposite argument. There were so many unanswered questions, so many holes in the preliminary field research, that it was not at all clear that the data from Minnesota was an accurate reflection of what was happening. Sessions's reservations arose mainly from what Dave Hoppe had reported earlier in the day, as well as from Martin Ouellet's findings—which Hoppe told me during a break appeared to be the "Canadian parallel."

In his presentation, Hoppe had explained that two things must be happening in Minnesota. At least. He talked about the kinds of deformities he'd recorded and how the magnitude of what he'd seen had convinced him that this was "a new, rapid-onset problem." But he also conceded that there was quite likely a natural cause for at least some amphibian deformities because of the long history of sporadic outbreaks. In 1950, Hoppe said, some 350 bullfrogs with multiple rear legs had been found in a wetland in Mississippi. None of them reached reproductive maturity and they were never seen there again. This, said Hoppe, seemed to be characteristic of the historical record. Deformed frogs—in particular frogs with multiple hind limbs—apparently did show up now and then for brief periods, without rhyme or reason, in what scientists call "stochastic" episodes, meaning accidental and random. What Hoppe was seeing now was different, he said. This necessarily meant that a new kind of deformities outbreak spreading across the state was superimposed upon a preexisting phenomonon. Presumably these different types of outbreaks would have different causes, but how could anyone begin to sort out one from the other? The historical record Hoppe had surveyed cast doubt on the significance of what was being seen in Minnesota, but only by degrees. Not absolutely. Somehow the deformities that were "natural" had to be sorted out from those that were not.

But how? The historical record pointed mainly to outbreaks of extra legs. Most of the recent reports in Minnesota and Canada documented missing legs. But there was overlap. Sessions had seen both phenotypes at Aptos, and both had also occurred at the Ney farm. Hoppe said he couldn't understand how either deformity would have been missed in leopard frogs before this, if in fact they occurred in the species with any frequency. The northern leopard frog, Hoppe said, had been harvested and studied for decades without any report of such deformities. He said that he and Bob McKinnell had together examined nearly 20,000 leopard frogs and that he'd never seen such limb abnormalities in the field. "Actually, I do have one missing limb recorded in my field notes," Hoppe said. "At the time I attributed it to an unsuccessful predation. But I'm reconsidering that now." The frogs Hoppe had found lately that were missing limbs had seemingly perfect mature skin growing over the place where the leg should have been. There was no evidence of scarring.

During the 1996 field season, Hoppe found deformed leopard frogs in four out of six of his most frequently visited wetlands—places where he'd collected hundreds of normal animals in previous years. But the most provocative data had come from CWB, where so many different species were involved and where the evidence pointed so strongly to something "nasty" in the water because of the elevated rates of abnormalities in the more aquatic frogs.

This wasn't happening in a city sewer or the spillway from a chemicals plant. It was happening in the heart of a supposed paradise. Apart from the herd of dairy cattle that each day walked down to the water just up the shoreline from the Bocks' there was no obvious environmental disturbance to pin the deformities on. "This site is essentially a small, pristine northern Minnesota lake," Hoppe said. The juxtaposition had cut to the bone of what many people already feared were the implications of the deformities outbreak. How could such frogs have come from such a lovely place?

The data from Minnesota in some respects seemed to be more rumor than data. Hoppe's reliance on his own experience was understandable—invaluable, really. But he couldn't be everywhere at once. Judy Helgen had said the evidence that the deformities were in fact everywhere in Minnesota was what had her the most "spooked." It also made fieldwork difficult. The search for reference sites where the frogs were normal, she said, had been frustrated time and again by the discovery of more deformed animals. Helgen recalled a visit to one wetland they believed was "clean" only to hear from local kids that they'd seen deformed frogs there. "Sure enough," said Helgen, "when we looked, there they were." This raised the question of what the normal, background rate of deformities was. Was some percentage of frogs *always* deformed and was this just now being learned because of the intense attention suddenly focused on the animals in the wake of the Ney discovery?

This was a question that would be asked over and over again in the coming months, and I could never quite see its logic. If you believed that more deformed frogs were being found only because more people were looking for them—and not because there really were more deformed frogs than usual—then you already assumed that the deformities were a normal occurrence.

On the other hand, if you believed any of the deformities were a new, abnormal phenomenon, then it was a real problem regardless of the intensity of the search. Since there was no way to know which of these two possibilities was correct without determining the actual cause of the deformities, you were left pretty much where you started. It didn't matter how hard anyone was looking for deformed frogs because it didn't shed any light on the cause. The argument was closed and circular.

Later that same day, at the end of the session, I met Helgen in the hotel bar. She was fifty-eight, prim, round-cheeked, articulate. Scientists, I was finding, were always good with words. Helgen ordered a martini. I had a beer and we talked for an hour or so. Helgen seemed more relaxed but no less beleaguered than she had been during her talk. The MPCA, she said, was working on the assumption that an environmental insult of some kind was causing the deformities. She could not believe this was any sort of natural event; her personal level of concern, she said, was "very high." Helgen said she strongly suspected farm chemicals were involved. Because of the constant introduction of new pesticide products into the market—and because of their heavy, widespread use throughout so much of Minnesota—it seemed reasonable to infer that some recently introduced active ingredient was involved. Or that some new formulation was interacting with other chemicals already in use. This would explain the sudden outbreak of deformities. She hinted that the map showing deformed frogs turning up "everywhere" in Minnesota might be misleading. The truth was that only two confirmed reports had come from nonfarm settings so far, and one of those looked questionable. Helgen also thought that for a variety of reasons, including the "political climate" in certain parts of the state, many farmers might be hesitant to report deformed frogs on their property.

I asked whether Helgen thought the frogs might be warning of a potential threat to human health—a concern that had already become such a staple of the discussion that I really meant it less as a question than as a request for elaboration. Helgen sipped her drink thoughtfully. Well, she said, that might not be the most constructive way to think about what was happening. Separating ecological well-being and human health doesn't always make sense, she said, and anyway, it was important to attach a significance to both. The distinction between the natural world and the human world was a false one. It should be enough just to know that a whole class of organisms was in peril, whether that indicated any direct risk to human beings or not. This was a point I would hear Helgen and many other biologists make again and again in the future. "We have to care about the biological community itself," she said. "This kind of disturbance is very troubling. But naturally we also have to be concerned about what it says about the ecosystem that we're living in."

Helgen said she didn't think concerns about a danger to humans were misplaced, but that they were still speculative. Some of the hardest questions she'd had to answer over the past year had come from people who were already scared. A couple phoned the MPCA and said they'd seen deformed frogs near a house they were hoping to buy. They wanted Helgen to advise them on whether to close on the deal. Another time, a pregnant woman who'd seen abnormal frogs on her property called wondering if it was safe for her to stay there. Helgen said she never knew how to respond to these inquiries. I asked her whether, given the explosive nature of the outbreak, there were plans to issue any kind of health advisory to the citizens of Minnesota.

"I don't really know," she said quietly. "But I will say this much. Sometime this year or next we're going to have to take a very hard look at that issue."

I tried to imagine what form such a notification might take in Minnesota, the "Land of 10,000 Lakes," as it had been advertised on license plates for decades. How could Judy Helgen, or anyone else, tell people that the ground and water all around them might be poisoned?

————————

And so an indistinct, slightly wobbly picture began to emerge over the course of two gray days in Duluth in the fall of 1996. Most people at the meeting thought that however the investigation was run in the future, the deformities problem had now been roughly delineated as a kind of dialectic oscillating between two possibilities: parasites and chemicals. Stan Sessions endorsed a "two-pronged approach" that would give equal weight to either cause. But everyone also agreed that other possible causes—ultraviolet radiation, acid rain, disease, or some unknown mix of factors—still had to be considered.

In a way, the action plan sounded to me like little more than a description of the original problem. The rules of engagement were simple. Everything still had to be considered. Deformities could be natural, a perfectly normal event that happened occasionally in habitats that are generally remote and unmonitored but were now being combed over by an army of researchers and volunteers. Or deformities could be a completely unnatural, extremely dangerous warning of an environmental hazard that would presently catch up with other living things, including us.

Or both.

When I caught up with Dave Hoppe during a break he cautioned against assuming that something that caused deformities in one ecosystem would necessarily do the same thing in a different setting in another part of the country and among different species. Parasites were being evaluated in specimens from Minnesota. They certainly appeared to be the explanation for what Sessions had seen in California, Hoppe said. But he was dubious about such a scenario playing out in the upper Midwest, where long, frigid winters and short, unpredictable summers forced amphibians to be flexible. The outbreak in Minnesota, he said, was widespread and involved numerous species that bred and developed on very different, highly variable timetables. Hoppe thought the coevolution of a snake-to-snail-to-frog-to-snake cycle would be a difficult trick in Minnesota wetlands, where frogs were opportunistic, often explosive breeders whose life histories were closely tied to the weather. In any event, it was far from proven that parasites could cause the sorts of deformities that constituted the bulk of the data from Minnesota—skin webbings, crooked legs, missing legs. "I guess parasites might be involved here somewhere," Hoppe said. "But I think if you're looking for a single root cause that's not going to be it."

At the close of the meeting, a handful of scientists gathered around the lectern. Tietge said he'd get out a written summary of the workshop and the proposals for future research as quickly as he could to all of the participants. I asked him how long he thought it would take to get to the bottom of what was happening. A long time, he said. You needed long-term trend data. You might be looking at a complex set of cofactors that would be very, very hard to work on in controlled laboratory conditions. These were questions that could take years to answer. I said that I could see that, but didn't it seem that the problem might demand a more urgent response? Tietge thought about this. I looked at Ken Muneoka, who was standing next to us. He looked back at Tietge and me for a moment, then reached up and vigorously scratched his head. I thought maybe I was asking a stupid question, but when I got to know him better later on I learned that this head scratching was an indication that Muneoka was thinking hard and was about to say something that was almost always worth listening to.

"Let's say I'm walking down the street and I see you coming toward me and I notice that you have only one arm," he said. "Now that's too bad for you, of course, but it doesn't alarm me because I can think of many reasonable and common explanations for a missing arm. But now let's say I see you coming toward me on the street and I notice that you have two normal arms—plus a third one projecting out from the middle of your abdomen. Now this worries me. Because without knowing another thing, without further evidence of any kind, I already know that something went way, way wrong with you."

Tietge nodded at this. I asked him how worried he thought we needed to be.

"If it turns out to be a chemical problem, then we'll have to do a risk evaluation for humans," he said. "After all, frogs are vertebrates and so are we."

I drove back to the Twin Cities that afternoon. The next day I wrote and filed a story on the conference for the *Post*. It ran on Monday, September 30. That night I turned on the news and was surprised to see Dan

Rather reading a story about deformed frogs while images of Cindy Reinitz and her students at the Ney farm played on the screen. The following day Rick Levey, a biologist with the Agency for Natural Resources in Middlebury, Vermont, received a phone call from a local citizen who said he'd read about the deformities in Minnesota and thought he seen some frogs like that during the summer near Lake Champlain. Lake Champlain, the sixth largest lake in the U.S., is 700 miles from Minnesota. It borders Vermont, New York, and Quebec province. On October 9, Levey and another researcher visited four sites along the eastern shore of Lake Champlain. They collected 230 leopard frogs. Almost 17 percent of the animals had limb deformities, mainly missing hind limbs.

8
Choosing Sides

PARASITES. CHEMICALS. THESE WEREN'T JUST TWO IDEAS ABOUT what could be causing deformities in frogs. They also represented two sharply divergent ways of thinking about the world. The scientific method begins with an inference—an educated guess—but this is inevitably biased by attitudes and beliefs. Some people see nature as infinite in its ability to produce unexpected effects and, conversely, in its capacity to limitlessly absorb any changes we force on it. Other people think it patently obvious that human beings are relentlessly modifying the environment and, in the process, altering the natural order of the world. Scientists are divided on this issue, too, but not equally.

Most of the many biologists I've met and talked with over the past several years believe there is no longer any place on Earth so free of human impacts that it could still be called *pristine*. With that belief comes the inchoate but growing conviction that chemicals, alone or in unimagined combinations, are undermining the biochemical machinery of all the life-forms that evolved in the absence of manmade chemicals. For a lot of biologists, the case against chemicals is self-evident. We swim in an ocean of manmade chemicals—somewhere between 60,000 and 100,000, maybe more, no one knows the actual number. We do know that an entire class of manmade chemicals—pesticides—are formulated to kill or suppress living organisms.

The fact that we appear to live quite well in this saturated environment can be taken as a sign that most chemicals are benign to us, at least in the doses we routinely encounter them. But chemicals are not universally harmless. And because there are so many of them, in so many places, in concentrations and combinations no one can even guess at, chemicals are under a perpetual cloud of suspicion. Among the effects chemicals are capable of causing in vertebrate animals is abnormal limb development—countless experiments have proved that. Chemicals are widely distributed across the landscape, and Martin Ouellet's work had already shown an association between agricultural chemicals and frog deformities in Canada. It wasn't hard to suspect that chemicals were involved in the deformities; it was hard not to suspect it.

The evidence that parasites could in some way alter limb development, meanwhile, was meager. Stan Sessions's field observations, plus his lone experiment using resin beads as stand-ins for parasitic cysts, were it. But parasites did offer a natural, less-alarming "ecological" explanation for limb abnormalities. This appealed to anyone who thought the chemical explanation needlessly alarmist. It was also refreshingly simple. By the time of the Duluth meeting it was generally understood that at least some deformities were a natural phenomenon. Why not all of them? Nobody had ever seen so many deformities across such vast areas before—but it was undeniable they had been seen now and then. And the world is a big place.

What I thought I'd heard Sessions say in Duluth, however hesitantly, was that it might be both parasites and chemicals—or anyway, that parasites didn't seem likely to explain everything. This view was agreeable to just about everybody. But the plain truth was that the two causes were never equally weighted in anyone's mind, and in the coming months biases and attitudes hardened into absolute convictions. About a year after the Duluth meeting I was talking to Joe Tietge about natural versus unnatural causes for deformities. I happened to mention that certain kinds of deformities and certain kinds of sites looked the same to me.

"What do you mean?" he said.

"Well, I think a leopard frog with a missing leg from Vermont or Canada looks pretty much like a leopard frog with a missing leg from Minnesota," I said. Bob McKinnell had remarked on this, too. When the Vermont researchers had sent him frogs from the Lake Champlain basin

McKinnell found the deformities identical to many he'd seen in Minnesota and told me he suspected they must have been caused by the same thing.

"What you're saying is that there's normally a lot of variation out there in nature, so when you see instances where there's consistency in a pattern it suggests a connection," Tietge said.

"Yes," I answered, wishing I'd actually put it like that.

"Well, I see your point," he said. "But I'd have to disagree. I don't think this is going to turn out to be just one thing. There's no silver bullet."

Parasites and chemicals were in an important way parallel hypotheses: Both causes were based on the idea that deformities were induced by altering the positional identities of cells within the developing limb. Cells that were lost and disoriented, whether by interference from chemicals or parasites, without proper instructions on how to build limbs proceeded to build them wrong.

But there the similarities stopped.

A chemical cause would be undeniable evidence of human modification of the environment. Parasites presented a more complex set of possibilities. If deformities were on the increase, as they appeared to be, was infection by parasites also more common, and if it so, why? Were parasites increasing as a result of some other factor? If Martin Ouellet was right that deformities were more common in wetlands exposed to pesticides, then why would parasites be more of a problem there? Could exposure to chemical contaminants weaken frogs and make them more susceptible to parasitism? These were hard questions, but parasites also held out the promise of more concrete answers, too. Parasites were a physical presence. They would be easier to isolate and identify than some unknown chemical contaminant. Sessions's work showed that parasites rode along with the frogs they deformed—they were a cause that stayed with the problem. If parasites were involved, it ought to be a snap to figure that out. A chemical, by contrast, would work invisibly, leaving only the abnormality as evidence of its contact with the frog.

Of course the biggest difference between chemicals and pesticides was what was at stake. A chemical cause of the deformities could pose a serious threat to other organisms, including humans. But a parasite with an aquatic life cycle that used larval amphibians as temporary hosts probably wouldn't threaten other species. It was all but impossible to imagine that parasites would have any impact at all on people.

I thought there was another dialectic to consider—another set of precariously balanced opposites—although for a long time nobody else ever mentioned it. What everyone in Duluth seemed to agree on—including, for now, Stan Sessions—was that a new kind of deformities outbreak had begun to occur more or less on top of whatever happened normally. The fact that most of the current reports dealt with missing legs, while most of the historical reports dealt with extra legs, hinted that it might be possible to separate the old syndrome from the new according to the *composition of the deformities*. Put another way, were there sites where deformities were "natural" and other sites where they were "unnatural"? And if a new epidemic of unnatural deformities was being superimposed on the natural ones, how could these be distinguished from one another? Just about everyone assumed that something like this was happening, but nobody ever tried to separate their observations into "old" and "new" categories. All the deformities were treated as a single phenomenon. Every report of malformed frogs was considered part of a new, emerging problem, even though this was clearly not possible.

Think of it this way. Picture a map of North America. Take a blue pencil and tap its point here and there across the map to leave a pattern of scattered blue dots. Say those are sites where deformities are occuring naturally. Now take a blue pen and tap its point across the map, mixing in with the old pattern. These dots represent new outbreaks of new types of deformities. Now you have a commingling of different kinds of points on the map. Some are ink. Some are pencil. Each represents a distinct phenomenon. But you can't tell which is which because they all look the same.

Of course, it was still possible that all the deformities were "natural." Although this seemed to be contrary to the experience of most of the seasoned field biologists who'd weighed in on the problem so far, in a purely objective, scientific context, the possibility could not be ignored. But in the real world, it had to be ignored. We don't live in a context of absolutes. We live in a fluid, sometimes dangerous context of day-to-day risks and uncertainties that we try to minimize even when we don't fully understand them. There are different ways to do this. One way is to assume the worst when there is doubt.

Scientists actually have a terminology for this kind of decision-making. In matters pertaining to public health and well-being, scientists recognize that they must sometimes choose betweeen making one of two kinds of mistakes. A "Type One" error is when you warn that something is dangerous that later turns out to be safe. Type One errors are simply false alarms. But a "Type Two" error is when you say something is safe that later turns out to be dangerous. Type One errors are embarrassing and may raise credibility issues. But people don't get hurt by them. Type Two errors are a tragedy. Nobody wants to make a Type Two error.

So assuming the worst in the deformities problem was easy. It was the only responsible thing to do.

The deformities outbreak at CWB suggested that it could happen anywhere—although "anywhere" in Minnesota is usually not very far from the kind of agricultural activity that Martin Ouellet believed was connected with the frog deformities in Canada. The subsequent wave of deformities reports from around the state only reinforced the idea that there was something out there, something in the water that posed a threat to frogs, and, by extension, to any other living creature that used those same waters. The possibility that Minnesota's 10,000 lakes might be unsafe to swim in or eat fish from was all but unthinkable—except among the handful of scientists who'd been in the field for two seasons now and who had begun to think exactly that. McKinnell had warned his colleagues that dose and exposure were primary considerations, and that people, unlike frogs, do not live and breed and rear their young in water. But this only seemed to add another building block to the foundation of what had become a tower of questions without answers. Through good planning and a little luck, Joe Tietge had managed to bring together a group of government and academic scientists who could begin to answer these questions, and when the Duluth meeting ended I thought that was what they would go right out and do.

I could not have been more mistaken.

If science as an intellectual discipline can be compared to a great city, then science as it is actually practiced in the everyday world must be said to be more like a collection of medieval fiefdoms. Agencies and independent academics live and work in walled enclaves, among which communication and cooperation is possible but often strained. In general, these principalities are in competition with one another—for resources and for the glory of being the first to discover and publish the data and conclusions that will become the basis for further inquiry by everyone else. Glory begets more resources, which beget further glory. And so on. At times this competition is fierce, bordering on open hostility. Scientists, it turns out, suffer the same human foibles as everyone else. They're driven by ambition, by money, by ego, and by the perverse impulse to succeed amid the wreckage of someone else's failure. For all its geniality, the Duluth meeting, I eventually came to realize, was an anomaly. Over the next three years, as I visited and talked with and got to know many of the original participants from Duluth, I learned that scientists are staunch individualists, apt to dislike one another.

This animus is only rarely personal, however. The knot of scientists you will find in the hotel bar at the end of the day at any conference is likely to be an exceedingly jovial, well-oiled gathering. What they don't like about one another only becomes evident when there are ideas in competition. Much as we might want to believe that scientists are open-minded and ruthlessly objective, the truth is otherwise. Scientists are as blinkered and stuck on their own pet theories about everything as anyone else. Scientific method is supposed to be about proposing testable hypotheses and then attempting to disprove them. It would be a much prettier process if it actually worked that way. More often, scientists propose hypotheses and defend them fiercely while trying to tear down everyone else's. The idea that participants at the Duluth meeting had coalesced around a common objective was fanciful; the mood lasted only as long as it took for people to pack up and head for home. As fall gave way to winter, the North American deformed frog investigation, such as it was, marched away from itself in several different directions.

In October 1996, the same month that deformed frogs were discovered in Vermont and right on the heels of the Duluth meeting, George Lucier got on a plane for St. Louis, where he was to give a speech to the Society of American Environmental Journalists at their annual convention. On the way he came across a story in *USA Today* about the deformities outbreak near Henderson. Lucier is the director of the Environmental Toxicology Program at the National Institute of Environmental Health Sciences, a federal agency headquartered on the shores of a quiet lake in Research Triangle Park, among the piney, rhododendron-laced woods of central North Carolina.

Lucier's is an obscure but lofty position in the government's hierarchy of human-health watchdog agencies. The NIEHS—it's one of the twenty-five National Institutes of Health—isn't widely known outside of scientific circles. But it's arguably the premiere human toxicology research lab in the world, and it is the principle player in the multiagency National Toxicology Program, which is yet another federal enterprise that investigates issues relating to chemicals and human health. Unlike the EPA, the NIEHS has no regulatory responsibilities—its mission is pure research. This makes the agency's mandate open-ended. As Lucier explained it to me, their job is to look into "situations" and "exposure circumstances," and to study "basic mechanisms."

Lucier, an eminent toxicologist in his own right, is a tall and sober man whose demeanor is about right for someone whose job is to ferret out deadly substances lurking in the chemical mix that is the modern world. He recalled for me, many months after that flight to St. Louis, that the news of the deformities had caught his attention as exactly the kind of issue his agency needed to be investigating. "I thought to myself, 'Gee, this is something that is of interest,'" he told me. Not exactly a lightning bolt of alarm, but enough to get things rolling.

Lucier could assign people to work on the deformities problem almost immediately, and that is exactly what he did. He really didn't see how he could do anything else; the situation literally demanded an intervention by federal toxicology officials. At the NIEHS it's routine to respond to a wide variety of requests to study potential chemical toxicants. Anyone—a doctor, another government agency, even a private citizen—can nominate a suspect chemical and have it considered for further investigation through the National Toxicology Program. Lucier thought the facts, slim as they were at that point, amounted to a kind of toxicological probable cause. "In my mind this is the sort of thing that the public pays their tax dollars for us to do," Lucier said. "We're not supposed to sit around in our offices and play with molecules. We're supposed to deal with real situations that people are concerned about. Well, it was clear that people were concerned about possible human health effects these deformities might be signaling in Minnesota, so that's our responsibility. To determine whether they are real concerns or whether they are not. I just viewed it as part of what we get paid to do."

"Did it look like a hard problem or a simple problem?" I asked.

"We get many different types of problems, so you have to put that into context," he said carefully. We were in his office in Research Triangle Park and he had gotten up from the conference table where we were talking to get something from his desk. He paused to look out the window. "I thought this was a relatively hard problem," he said finally.

As soon as he got back to North Carolina Lucier held a series of meet-ings with his staff, trying to determine what his agency should do. After talking with Judy Helgen and holding a number of brainstorming sessions around the conference table in his office, Lucier sent a five-person group to Minnesota in December. The team was headed by one of Lucier's deputies, Mike Shelby, and by a quiet fifty-year-old toxicologist named Jim Burkhart, whom Lucier and Shelby had picked to lead a collaboration with the MPCA, if one could be arranged.

Shelby and Burkhart went to St. Paul with an offer the Minnesota researchers couldn't refuse: If the MPCA would make a formal request for assistance on the frog problem, the NIEHS stood ready to help. Nobody else had put it quite that way before—the only help Helgen had gotten so far was a little money from the legislature, which she had used in turn to pay for Dave Hoppe and Bob McKinnell and for a smidgeon of water chemistry reports from several private laboratories. Burkhart said his agency was prepared to initiate a "rapid, direct" inquiry into the frog prob-lem to see if anything "really gross" could be determined right away in terms of a cause. Burkhart told me later that a quick response was part of the pitch, but that he figured this would ultimately turn into a much longer-term project. He said he didn't discuss finances directly with the MPCA, but that it was clear to everyone that NIEHS was going to pick up the tab for just about everything but field collections, which would still be the MPCA's job. And because the project was to be internally funded at NIEHS, there would be more money if it were needed in the future. This was welcome news. Helgen's in-house research capabilities at the MPCA were somewhere between slight and nonexistent. Her corner of the water quality "lab" amounted to not much more than a couple of tables, a stain-less-steel sink, and a broom closet. Burkhart's proposal to not only do all the water testing but to pay for it as well looked heaven-sent.

———————————

Like Joe Tietge up in Duluth, Burkhart was prospecting for business. He, too, was part of a newly created team for which the frog problem had loomed as a sudden target of opportunity. Burkhart was the leader of something called the Alternative Systems Group, a research unit that had been formed in 1995 to respond to the growing complaint that the field of toxicology relied too heavily on a handful of familiar laboratory animals as test organisms. The bioassay is the backbone of toxicology, but every "model" species has its limits. Basically, the concern had become that mice and rats, plus the handful of other species routinely used in testing chemicals for toxicity, didn't provide a diverse phylogenetic lineup. Although many biological processes are conserved among different species, not every animal responds identically to toxicants—as the thalidomide tragedy so graphically demonstrated—and the search for a broader spectrum of bioassays had been joined.

Somehow this need for new test species had to be reconciled with a different new concern. Like every other research facility that used animals in experiments, the NIEHS was under mounting pressure from the animal-rights movement to cut back on its use of live bioassays. In 1993, Congress passed legislation that required the NIEHS to develop new model species protocols and at the same time to invent "alternative methods that can reduce or eliminate the use of animals in acute or chronic saftey testing."

Burkhart interpreted these seemingly contradictory orders as a loose directive simply to do things differently. He was particularly interested in developing methods for monitoring wild animals as bioindicators of potentially toxic situations in the field. When I talked with him about this he explained that lab mice and rats typically came from commercially reared strains that had been isolated from the natural environment for many years. In his view, wild animals were a more relevant ecological barometer, a useful kind of natural bioassay if you could monitor their status where they lived.

Native species could also be brought into the lab—which might be very important if you were looking for a specific toxicological "endpoint" in a particular organism. Anybody who wanted to figure out what caused a leopard frog to be deformed would, sooner or later, have to be able to elicit the same kinds of abnormalities in a leopard frog under controlled laboratory conditions. Other types of test organisms were also under development. Fish, for example, could be reared in large numbers—something that could be important if you were looking for a developmental effect that only arose in a small percentage of animals. A year after starting work on the frog problem, Burkhart summed up his thoughts on all of this in a paper he coauthored with Henry Gardner: "Given the reality that certain contaminants in the environment threaten the resident organisms, our challenge is not whether non-mammalian and environmental sentinel species data can be used to evaluate human health implications, but how."

In Burkhart's thinking, all of these considerations merged in an open-ended mission to find new ways of doing toxicology. "Alternative is just a buzzword," he told me. "It means not a mouse. Not a rat."

Even so, Burkhart told me the frog problem was a fairly significant departure for the NIEHS. The agency had more often been involved in ecological
problems only through "extramural" funding of outside researchers. But
the frog investigation was going to be run right out of Burkhart's office.
This did not escape notice a few blocks away in Research Triangle Park,
over at the EPA's National Health and Environmental Effects Research
Laboratory—the mother ship for all of the EPA's far-flung research facilities, including Duluth. For years, the EPA had enjoyed a virtual monopoly
in the investigation of toxicology issues relating to wildlife. But in the
1990s, it became clear that NIEHS was increasingly interested in the connection between environment and human health, and that the agency no
longer considered wildlife toxicology a separate category. This was especially vexing to people at the EPA, who saw themselves already disadvan-
taged relative to the NIEHS when it came to maintaining programs or
expanding into new areas. Because it is a regulatory agency, the EPA is
strictly limited in its ability to lobby Congress for support. The NIEHS is not
under similar constraints. Whenever the NIEHS got something new
going—like the Alternate Systems Group—people at the EPA felt it had
come at their expense. The NIEHS helped fund work on cancer in fish and
also got involved in the investigation of the *Pfiesteria* outbreak in eastern
coastal waters, where the toxic single-celled organism that had killed millions of fish and shellfish also appeared capable of making people sick.
Burkhart's arrival in St. Paul within two months of the Duluth meeting—
nobody from the NIEHS had even been invited to Duluth—was thus a
small adventure into territory the EPA had already begun staking out for
itself.

Burkhart insisted to me that he never saw his agency and the EPA as
being in competition. Certainly that wasn't his style, which was strictly
laid-back. An army brat, he'd grown up in different places. He went to college in Missouri in the 1960s and started out studying biology and chemistry, but got caught up in the cultural vortex of the times. Burkhart ended
up earning a degree in political science. Then a tour in Vietnam as an army
corpsman caused him to reevaluate his interests once more. "I think that
experience sort of took the politics out of me," he told me. "After Vietnam
I felt like politics was not such an important thing. It just seemed kind of
trivial." When he got out of the army, he went back into science. One of
the things I liked about Burkhart was his sense of limits—he thinks in

broad terms about how to approach problems but concedes that getting all the answers is not always possible. It's a philosophy of interactions, a way of looking at life I've run into time and again with biologists. Biology is complex. Factors collide and impinge upon one another. Nothing is simple. On a couple of occasions Burkhart and I talked about how much of what happens to us in life—and death—is prefigured by our genes and by what happens to us during development.

"Do you think we're programmed to die at a certain age?" I asked him one time.

He laughed. "I don't know if you're programmed or not," he said. "What it could be is that you can eat and drink and smoke and do all this sort of stuff and maybe it really doesn't make that much difference. Unless you do it before you're born."

Anyway, Burkhart was a flexible man, and that's what the frog situation called for at the moment. Somebody had to start somewhere. Why not us, Burkhart reasoned, and why not begin with the water. "When I first read about these deformed frogs," Burkhart told me, "what it said to me was that it was likely there was something in the water that was affecting a primary developmental pathway. I looked at the research in both frogs and mammals on effects of exposure to different kinds of hormones or retinoids. Of course, that was immediate. But there is also the reality that the environment is a very, very complicated setting, especially when you're talking about development and when you're talking about a frog that has a very permeable membrane. It seemed very important for us to ask what was actually in the environment where this was happening. Because we didn't know."

Burkhart didn't think the MPCA had learned much in their first full season of investigation. One of the things he did in St. Paul was sit in on a conference call with Stephen Goldberg, a parasitologist at Whittier College in California who'd been brought into the investigation by Bob McKinnell the previous year to evaluate frog specimens from Minnesota. Burkhart told me he wanted to know whether parasites were a possible cause of the deformities, and what he'd heard in St. Paul convinced him that parasites could not possibly explain everything that was being seen in Minnesota. Goldberg, who had worked in parasitology and herpetology for thirty years and coauthored more than one hundred papers on parasites in reptiles and amphibians, said he'd never seen anything like what was happening in

Minnesota—but that he was sure it wasn't all the result of parasites. Goldberg had already examined hundreds of deformed frogs from Minnesota and he told Burkhart he couldn't rule out parasites in every case, but he was certain that something else had to be causing at least some of the deformities because he was seeing abnormal frogs that were free of the parasitic cysts that Stan Sessions had proposed as a cause.

Burkhart told me the information the MPCA had at that point was mostly "little pieces" like that. The data didn't yet point in any "definite direction." This, he said, wasn't terribly surprising. When you're dealing with an environmental problem like this, he told me, it's important to limit the data you're trying to understand. "It's easy to get more information than you're able to interpret," he said. The MPCA's decision to attack the problem in small, manageable increments seemed like the right idea to Burkhart. The fact that the Minnesota researchers had been overwhelmed by the number of deformities reports that year in spite of their best efforts had added a completely understandable level of confusion to the investigation.

What bothered both Burkhart and Lucier, though, was the working assumption that had guided the Minnesota team from the beginning. Nobody at the NIEHS thought it would be likely that anyone could find a chemical cause for the deformities merely by looking closely for some common agent that was present everywhere the abnormalities occurred. Lucier was struck by how "strong" the response was in the deformities— that is, by how site-specific these outbreaks were in a state that was not generally regarded as being highly polluted. But he didn't think it would be easy to link all the sites to one another with a common agent that would turn out to be the ultimate cause in every instance. This was exactly the same assessment that Joe Tietge had made. When I had asked him about this almost a year and a half later, Lucier told me that the MPCA's hope of finding a correlation between the deformities and a particular chemical or even a family of chemicals was probably wishful thinking. "We talked a lot about that when we began thinking about the studies and what we should be doing in the way of water analysis," he said. "My feeling was that I didn't think anything would fall out of that approach."

"Why not?" I said.

"I just didn't think it would," he said. "But your gut feelings can be wrong and it's important to have baseline data, because no matter what happens other people are going to say it's this chemical or that chemical. When we have data on the chemicals that are present then we can respond. So it was important for us to develop a database of what was in the water, even though I thought it was unlikely that something could fall out and we'd have our answer. I wasn't certain that it wouldn't happen. I recognized the possibility existed. But if you had asked me to bet on it, I'd have bet that it wouldn't happen that way. Less than a fifty-fifty chance."

"And so far that's been the case?" I said.

"Yeah."

Burkhart assumed the same thing, too. He had devised an approach that he regarded as simpler and more likely to succeed than a blind search for a common agent. When I talked with him by phone early in the winter of 1997, he explained that the NIEHS hoped to begin in the coming summer by answering a single question: Did wetlands that produced deformed frogs have a chemical load that was different from wetlands where the frogs were normal?

Getting the answer would involve something called a "paired study." What Burkhart needed from Helgen were a handful of sites where there were deformities—plus matching sites nearby for each one that was similar but where the deformities didn't occur. Rather than looking for *similarities* between places with deformities, Burkhart said the NIEHS would be looking for *differences* between affected sites and their associated controls. In short, he wanted to answer one of the most perplexing problems that had been identified in the deformities investigation. Why was it that one body of water should be a problem while another one close by was not?

———————

There was, however, an enormous assumption hiding within Burkhart's seemingly straightforward research proposal. The implicit idea was that if he could find a chemical "load" in ponds where the deformities occurred he could then isolate and identify a specific compound (or compounds) that actually caused the abnormalities. This was a tall order. The problem with this approach—as the EPA in Duluth had learned time and again—is that a guilty compound among the many compounds that are almost certainly going to turn up in an environmental sample has many places to hide. As Joe Tietge liked to say, you only find what you look for in a complex mixture. Burkhart was proposing to look for literally everything.

As daunting as this prospect was, however, it was not in any absolute sense impossible. The basic technique for isolating bioactive compounds in complex mixtures is a process called fractionation. Fractionation is a multistage process of chemical simplification, in which complex mixtures can be reduced to smaller and smaller subsets of similar chemicals. Essentially a sample is divided and redivided over and over again, until a complex mixture has been reduced to many discrete and much less complex mixtures. Within these "fractions" specific compounds can be identified, usually by determining their mass or by measuring the way they absorb light. Many compounds yield characteristic "signatures" by these methods; unfortunately the signatures for many more are unknown.

Fractions can also be tested biologically. This is what the EPA called the "effects-based" approach. Fractions can be tested individually in bioassays to determine which ones produce a biological response. It's a powerful process of elimination. Imagine, for example, that you have a complex mixture that contains 1,000 compounds. Assume that one of those compounds is causing deformities in frogs. That means that before you do anything you know that there are 999 dead ends in the mixture.

Now suppose you can fractionate the original mixture down to ten fractions, each one with just a hundred compounds in it, and you're only one step away from a major leap forward. Test each fraction on, say, leopard frogs, to see what happens. Only one fraction should result in limb deformities. Immediately, you are done looking at the 900 compounds in the "inactive" fractions that have been eliminated in a single step. It's not nearly so simple in practice, but that's more or less the idea.

Burkhart did not quite have the right bioassay at his disposal—ideally such a test would use one of the native frog species that was suffering high rates of deformities in the field, like leopard frogs or mink frogs. But he had something he felt was close enough. Even though it used an exotic species and involved a different endpoint from anything being reported in the wild, it was a test he was eager to do more work with in the Alternate Systems Group. It was called FETAX.

The acronym stands for "frog embryo teratogenesis assay *Xenopus.*" *Xenopus* is short for *Xenopus laevis*, commonly known as the African clawed frog. As the common name implies, *Xenopus* is not native to North America. But it's a popular pet, and enough African clawed frogs have either escaped or been released in Southern California to establish a number of wild populations in several counties and to be listed in field guides for North American amphibians. People like African clawed frogs as pets because they're so easy to care for. The animal is totally aquatic, spending most of its time submerged. It doesn't do much except reproduce and eat, and *Xenopus* can live contentedly on a diet of ordinary goldfish food. For all of those reasons, *Xenopus* is also a favorite with scientists. In fact, it's the only frog species in regular use in laboratory experiments.

FETAX was developed out of work originally done in the biology division at the Oak Ridge National Laboratory in the 1970s, when the Department of Energy and several branches of the military were looking for a fast, high-volume bioassay that could be used to test for environmental contaminants in complex mixtures. In the mid-1980s, several scientists at Oklahoma State University began working with FETAX and eventually took it into the private sector as part of the product line at an environmental consulting company called the Stover Group, located in Stillwater, Oklahoma. Burkhart had been collaborating for some time with researchers at Stover and also with another FETAX expert, his colleague Henry Gardner, who worked at the U.S. Army's Environmental Health Research Detachment at Fort Detrick in Maryland. They were continuing to refine and "validate" the assay in a variety of applications. Burkhart saw FETAX as exactly the kind of technique he was supposed to be developing as an alternative to traditional mouse and rat experiments.

———————

On the many occasions when I talked with him, Burkhart seemed resigned to the things that inevitably go wrong or just don't pan out in toxicological studies. You work hard to show an effect in the lab and you make a prediction about what that might mean in the world at large—but you never know that you're right until a negative outcome tells you one way or the other. Lab tests can suggest that a given compound is toxic to humans, or that it's safe, but you don't do tests with humans. They do it to themselves. And then you find out. Nobody ever "tested" thalidomide on people.

Burkhart had to be philosophical about the shortcomings of his profession. FETAX, though reasonably well established as a standardized test procedure, was still a crapshoot. All bioassays are gambles. People aren't frogs. Burkhart looked past the uncertainties and went to St. Paul and met with Judy Helgen. He told her about the paired study and the chemical loads and the fractionation, and he allowed it all to register. The NIEHS, without saying as much, was actually prepared to fund all this. All the MPCA had to bring to the table were deformed frogs, a lot of water, and a handful of "clean" sites that could serve as controls. How hard could that be? Down in Research Triangle Park they began drawing up a plan and a budget. The NIEHS would spend $600,000 on the deformities problem. For starters.

Back at the EPA in Duluth, Joe Tietge was busy with a different research agenda. The NIEHS was preparing to evaluate thousands of compounds in search of whatever was causing the deformities. The EPA was going to test just one: methoprene.

Methoprene is the active ingredient in a host of commonly used insecticides. It was first registered with the EPA in 1975 and has since been incorporated into a diverse line of pest-control products used by farmers, pet owners, and the government. Developed by Sandoz, a giant chemicals conglomerate, methoprene is currently manufactured in Texas and marketed by an Illinois-based company, Wellmark International. Wellmark is a pest-control spin-off that was created following a merger between Sandoz and another chemicals giant, Ciba-Geigy.

Methoprene is everywhere. It's in many over-the-counter products used for flea and tick control in household pets—dog and cat collars, dips, shampoos, sprays, and aerosol bombs. It's also sometimes added to cattle feed—not to treat cows directly, but to treat what comes out of cows. Methoprene is stable in the cow's digestive track and is "passed through" the animal and out with the manure, where it kicks in as a fly control. Methoprene is also used in a popular mosquito-control product called Altosid.

Methoprene was invented as a "safe" replacement for traditional nerve toxins like DDT and chlordane, chlorinated hydrocarbons that were banned for use in the United States in the early 1970s. The supposed safety of methoprene lies in its mode of action as a so-called "insect growth regulator." Methoprene mimics the hormones that sustain the juvenile stage in an insect's life cycle. It doesn't kill adult insects, rather, it keeps larval insects from maturing and becoming sexually reproductive. And because most insects have a short life cycle, even a few days of arrested development can effectively suppress a population as the adults die off and subsequent generations fail to mature.

It was long assumed that methoprene is harmless to animals that don't use hormones to promote juvenile characteristics. This would include all vertebrates. Methoprene was also thought doubly safe because it breaks down quickly in the environment. In the world of pesticides, specificity and rapid breakdown are extremely desirable features. But the EPA felt it had reason to think that methoprene might be a problem on both counts.

The first hint of a methoprene connection came in the spring of 1996, when Joe Tietge heard from an old friend named Pat Schoff. Schoff was a forty-three-year-old developmental biologist who worked nearby at the Natural Resources Research Institute, which was affiliated with the University of Minnesota, Duluth. Schoff and Tietge had known each other at the University of Wyoming, where Tietge had been an undergraduate while Schoff was working toward a Ph.D. About the same time that Tietge began looking into the frog problem, Schoff, who was prospecting for grant money, had gotten interested in how developmental abnormalities can arise as a result of exposure to xenobiotic chemicals, particularly those that interfere with hormone function—so-called endocrine disruptors.

The endocrine system is an integral part of the makeup of complex, multicelled animals. It is composed of the specialized glands and cells that produce hormones—powerful chemical messengers—that regulate development, metabolism, and reproduction. The thyroid gland, for example, secretes hormones that set and maintain the body's metabolic thermostat. In men, the testes secrete steroid hormones that promote sperm production and establish male sex characteristics. Similarly, the ovaries in women produce estrogen, a hormone crucial to the maintenance of the uterus and the establishment of female sex characteristics. There are more than fifty known human hormones. Most, though not all, are distributed throughout the body by the bloodstream.

In the endocrine system, a little goes a long way. Hormones are extremely potent, even at very low concentrations. Their receptors—the specialized cell regions that bind to hormones and activate genes—are likewise highly sensitive. And therein lies what many scientists believe is a profound complication for organisms living in the modern chemical age. The finely tuned sensitivities of the endocrine system, which took millions of years to evolve, are now being exposed to tens of thousands of new chemical compounds, most of which have come into existence in only the past fifty years.

Pat Schoff thought there was a possibility that the frog deformities in Minnesota had been caused by an interaction between the animals' endocrine function and some chemical in the environment that mimicked a hormone. When he saw photographs in the newspaper of the deformed frogs from the Ney pond in the fall of 1995, Schoff felt certain he'd seen those same kinds of leg abnormalities before—in books and articles discussing the role of retinoic acid in limb development.

Schoff also knew that retinoic acid was involved in many phases of development in vertebrates, including the formation of parts of the nervous system and skeleton. He knew, too, that retinoic acid was a proven teratogen—a compound that can cause birth defects. Too much or too little retinoic acid at the wrong time or in the wrong place had been demonstrated to cause developmental abnormalities in every vertebrate species tested—including humans, who became accidental test subjects through exposure to the acne treatment Accutane.

Accutane, which uses a derivative of retinoic acid as its active ingredient, was introduced in the early 1980s with a warning that it should not be used by pregnant women. It was used anyway, either by women who were already pregnant without knowing it or who became pregnant unintentionally during their treatment. In one study of women who decided to continue their pregnancies after exposure, about half miscarried or had babies with a variety of congenital birth defects—including abnormal jaws, missing or defective ears, heart abnormalities, and cleft palates. This really wasn't too surprising. Scientists have known since the early 1950s that vitamin A—the parent compound for retinoic acid—can by itself cause birth defects.

Schoff was the first person to look at the frogs from the Ney farm and propose a link to retinoic acid. He phoned Judy Helgen and told her he suspected that retinoic acid or some "retinoid" analog might be involved in the deformities. He even proposed a research partnership, asking that Helgen include him in her upcoming request for funding to the Minnesota legislature. Helgen, he recalled later, showed zero interest. At one point Schoff drove to St. Paul, to a seminar Helgen and McKinnell gave at the university on the deformities problem and introduced himself after the program. He explained again that because retinoic acid caused abnormal limb development in laboratory experiments a retinoid in the environment could do the same to wild frogs. Again, Helgen disagreed. Schoff told me later that he didn't think Helgen had much of a grasp of developmental biology. "She just didn't get it," he said.

Frustrated, Schoff phoned Joe Tietge—who did get it. Tietge asked Schoff to give a talk on retinoic acid to the EPA. In the course of his presentation, Schoff mentioned that he'd been looking for a candidate as the mystery retinoid in the environment—and he'd found one. It was the insecticide methoprene, which in addition to being hormone-based had a molecular structure eerily similar to retinoic acid.

Tietge made a mental note to get someone to come to the Duluth meeting to talk about the role of retinoic acid in limb development. But even before he'd gotten in touch with Ken Muneoka, Tietge got a shock. When he checked to see if there was anything on methoprene in the scientific literature, he found that tests of the pure compound on larval amphibians indicated that it had *no developmental effects* on them. But a researcher testing an out-of-date sample of methoprene had accidentally discovered that one of its breakdown products, methoprene acid, activates one type of retinoic acid receptor. Months later, during his talk at the Duluth meeting, Ken Muneoka had shown the audience a slide of a molecule he said might be an environmental retinoid worth looking into. The molecule in the picture was methoprene.

Gary Ankley told me later that he and Tietge were "flabbergasted" when Muneoka brought up methoprene seemingly out of nowhere. Tietge said they never once talked with Muneoka about methoprene before the meeting, although by that time they were already thinking about how to test it experimentally to see if they could induce abnormal limbs in frogs. Methoprene seemed to be forcing itself on the investigators. "When Ken mentioned methoprene it was a pretty convincing coincidence," Tietge said. "Kind of a magic moment."

The Biggest Experiment Ever

TIETGE'S PLAN WAS TO TEST METHOPRENE IN A BIOASSAY USING the northern leopard frog, *Rana pipiens*. It wouldn't be easy. Nobody at the lab had ever worked with a native frog species before, and to do the experiment they would need to force-breed females to obtain eggs so they could simulate the exposures that would occur in the wild. Meanwhile Tietge was busy with something else—organizing another meeting. This one was going to take place the following spring, in April, at Shenandoah National Park.

Tietge wanted to build on what had come out of the first workshop in Duluth, but in most respects the Shenandoah meeting would be a repeat of Duluth. Its real purpose was to go over essentially the same information again—the list of participants would virtually identical to Duluth's—but for a different audience, specifically for scientists from the EPA in Research Triangle Park who would be signing off on Tietge's work. Tietge thought they should hear firsthand from the experts.

The key member of this new audience was going to be Gilman Veith, the EPA's associate director for ecology. Everything that went on in Duluth was ultimately under Veith's supervision. Veith himself had worked in Duluth before moving up in the agency and had had a celebrated career in toxicology when he was in Minnesota. When I phoned him that winter, Veith said he already envisioned the EPA taking the "lead role" in the frog investigation. Wildlife problems weren't ordinarily an EPA concern, but the possiblity that disease, chemical contaminants, or increasing ultraviolet radiation might be involved had really gotten their attention. Veith said they would proceed, however, with caution. The reports of deformities outbreaks from one end of Minnesota to the other, and now across the northern tier of states and into Canada were riveting, but did they really say what they seemed to say, that an abrupt, unprecedented epidemic was under way?

"There are so many plausible explanations for what is causing these deformities that we need to be a little careful," Veith said. "We need to find out for sure what is happening first."

This general note of uncertainty seemed to be in the air over the whole frog investigation that winter. Judy Helgen was distressed when a retired newspaperman from Grand Marais named Duane C. Honsowetz wrote an opinion piece for the Minneapolis *Star Tribune* accusing the MPCA of wasting taxpayers' money on what looked to him like a quixotic search for a chemical cause of the deformities. The agency, he wrote, had obviously dismissed a perfectly obvious natural explanation for the frog problem— parasites—because it was "politically correct" to assume that chemicals were the cause. Honsowetz thought it was more likely that Mother Nature was to blame and said he couldn't tell if what the MPCA was doing was "legitimate research" or was actually the "Big Frog Caper." The piece ran under a large reprint of a picture already made famous on the Internet— Judy Helgen's shot of a leopard frog lying on its back with two pale, under-sized legs sticking out of its pelvic area between the primary hind limbs.

Helgen hadn't overlooked parasites at all, of course. She and McKinnell had started sending frogs out to Stephen Goldberg at Whittier early in their investigation, and Goldberg continued to report that there was no connection between the deformities in Minnesota and encysted trematodes or any other kind of parasites. Helgen told me later that Goldberg was so certain parasites weren't involved that he had promised to "eat his shirt" if anyone could show that parasites were causing deformities in the populations that he'd examined in Minnesota.

When I spoke with Goldberg some time later about his initial evaluation he was still quite positive that there wasn't any clear link between encysted trematodes of the sort Sessions had found at Aptos, California, and the deformities in Minnesota. His dissections had been complete and utterly thorough. The results, he said, were not open to alternative interpretations. "I found some association between parasites and deformities," Goldberg told me, "but nothing definitive. Certainly there were not concentrations of cysts around the sites of the deformities. Most important, we examined a large number of very grossly deformed frogs that did not have trematode cysts of any kind." In fact, Goldberg's data showed that nearly half of the deformed frogs he examined were clear of parasitic cysts.

Goldberg also wanted me to know that in his expert opinion—based on decades of research on parasites and amphibian hosts—he regarded the deformities as being unprecedented, notwithstanding the historical record of occasional outbreaks. "I've been getting my hands wet handling frog specimens in museum collections for forty years," Goldberg huffed. "You do not see examples of these kinds of gross deformities in those collections, not even in California. I've also looked at natural frog populations for many years. I've never seen anything like it. You know what those frogs from Minnesota looked like?"

I had a hunch I did.

"What did they look like to you?" I said.

"Well, they reminded me of those thalidomide babies," he said.

The MPCA had aquired another new partner that was working on parasites—along with just about every other conceivable cause for the deformities. It was the National Wildlife Health Center in Madison, Wisconsin. The center is part of the sprawling bureaucratic web of wildlife and research agencies that are part of the U.S. Geological Survey, which in turn is part of the Department of the Interior. Their main mission is to evaluate the cause behind any wild animal that is found dead or dying. The center tests for wildlife diseases and also does forensic work in necropsying animals that have been killed under suspicious circumstances. Madison gets a little of everything. Sick birds. Dead fish. Endangered species that have been shot. Deformed frogs.

The lead investigator in Madison on the frog problem was biologist Kathy Converse, who specializes in wildlife diseases. I called Converse that winter. She was extremely pleasant and quite firm in her conviction that there was absolutely nothing "natural" about the malformations in the frogs in Minnesota or in other places where limb deformities were being reported. She thought whatever was happening had begun only recently. "Look," she said, "there are a lot of guys out there like Dave Hoppe and Bob McKinnell. They haven't seen this before, and it's not a phenomenon that would have been missed."

Converse said her lab had evaluated about seventy frogs from Minnesota, and she herself had collected some of them between July and September. All were leopard frogs. Converse said they had found pretty much the whole nauseating array of deformities, except for extra limbs. This already seemed to be the general pattern. There was no evidence of either bacterial or viral infections other than what was normally found in frogs, although she said sometimes infection can cause a problem in an amphibian without leaving any evidence of what it was. She said they were convinced that the many missing or partial limbs they saw were not the result of trauma. She didn't think anything bit them off. She said she did not believe the world was suddenly full of incompetent snapping turtles or any other kind of predator that was going after frogs and coming away with only a piece of leg. Converse told me that a number of internal abnormalities in the skeletal and soft-tissue structures had been discovered. They found parasites, too, including what appeared to be encysted trematodes, though no identification had been possible.

But just as Goldberg had seen, there didn't appear to be any association between the presence or absence of parasites and what kind of frog they were looking at. Converse said they sometimes found parasites in normal frogs and sometimes didn't see any at all in frogs that had abnormal limbs. "As far as we can tell at this point, there's no correlation."

Like a lot of other researchers, Converse said she thought it made more sense to assume there could be more than a single cause of the deformities, maybe several. It could be parasites in one place and a chemical somewhere else, she said. The fact that deformed frogs were being discovered across such a wide geographical area seemed to argue against any one simple explanation. Converse said all of it was a worry.

"Are you concerned about what any of those causes might mean?" I asked.

"Sure," she said matter-of-factly. "This is a problem. It's a real problem. There's a much higher incidence here than we would ever expect to see in any kind of natural occurrence. It's not like we're collecting three hundred frogs and finding one little deformity."

On January 14, just weeks after the NIEHS visit to St. Paul, MPCA commissioner Peder Larson found himself under attack by a group of state legislators who were touring the agency. The legislators—among them Willard Munger, the chairman of the State House Environment and Natural Resources Committee who had sponsored the initial funding for the investigation—demanded an explanation of a story that had appeared in the Minneapolis *Star Tribune* earlier in the day, which reported that the agency was backing out of the frog inquiry.

The follow-up story, written by the *Star Tribune*'s veteran environmental affairs reporter Tom Meersman, described the exchange between Larson and the legislators as "angry." Senior management in the agency, including Larson, had overruled Judy Helgen's request to continue her work and would not ask the legislature to fund frog research in the coming season, Meersman reported. According to Larson and other managers at the MPCA, the agency had determined that the frog problem was beyond their capabilities and would be better handled by federal research agencies. Duane Anderson, Judy Helgen's section manager, told the *Star Tribune* there were "too many sites with way too many frogs" for the agency to deal with, and that trying to find an answer to the problem would "take far more time and effort" than the MPCA could possibly commit to such a project. As bad as that sounded, Anderson made it worse. He went on to suggest that the agency's management was not confident it could deal with the consequences if the investigation were to find a link between the deformities and farm chemicals.

"If the determination was that the cause was pesticides," Anderson said, "the pesticide industry would expect us to have that conclusion nailed nine ways from Sunday. We're not in a position to do that."

The MPCA's sudden change of heart—and their strange, befuddled defense of it—seemed to have been invented almost out of thin air. But the agency rallied around its newfound position. Two days after Larson's discussion with the legislators, Meersman filed another story in which Larson reaffirmed his contention that the frog problem was really outside the purview of the MPCA. "We are a regulatory institution, not a research institution," Larson told the *Star Tribune*. "We're not going to have an impact, in my mind, by putting $100,000 or $150,000 into the direct research. That's a university's job. We'll support that." Of course, that level of funding was assumed when Helgen had talked with Burkhart about supplying him with deformities data and samples from the field. The MPCA wasn't supposed to *solve* anything with that kind of money.

Not surprisingly, the agency's decision, if that's what it was, didn't please anyone. Bob McKinnell, with characteristic disdain, fumed on the record. "Inasmuch as frogs seem to be profoundly affected it's appropriate for the state to maintain a strong interest in the abnormalities," he told Meersman. "But I'm not going to tell them what to do." Joe Tietge was likewise caught completely off-guard. When I called him he seemed perplexed by the news that the MPCA wanted out. He told me it wasn't realistic for the state to think it could hand off to the federal government on this sort of problem. "The EPA can't do everything," he said. "It's not going to help for the state to withdraw. It's going to hurt. There has to be a collaboration."

Willard Munger and other legislators echoed Tietge's mystifcation at the turn things had taken at the agency. When I called Munger a few days later, he was seething. He'd just gotten a long e-mail message from Cindy Reinitz begging him to intervene. Reinitz wrote that she was "discouraged and disheartened" at the news that the MPCA was pulling out of the frog investigation. She said as a teacher she was contantly battling complacency among her students, and now here was the State of Minnesota reinforcing the "It's not my problem" attitude. "What does this say to the thousands of schoolchildren in our state as well as the nation who have been actively following this issue?" Reinitz demanded in her note.

Munger didn't need any convincing. He was every bit as mad as Reinitz and he wasn't going to stand by and watch the problem be ignored. "I'm not satisfied," he told me. "Not at all. I think they're running out on their responsibility." Munger said he wouldn't tolerate this kind of "buck passing" to the federal government. He planned to hold hearings. And he really wanted to know just what the MPCA was afraid of. Minnesota is an intensely agricultural state—was the agency really worried about where this investigation might ultimately lead? "I think they are afraid of moving ahead because they might step on somebody's toes," Munger said. "I think they're afraid this problem may ultimately involve pollution caused by pesticides."

The MPCA seemed to know it was in a fundamentally indefensible position. Their contention that the frog problem was too big for them to handle only convinced everyone else that the situation was desperate. Officials at the MPCA had tried to use language that only suggested they would not be asking the legislature for "additional funding." But the reality was that without a funding request for the upcoming season there would be no money for the frog program. When I talked with Duane Anderson he told me the agency's "reevaluation" of its role in the investigation began when the problem itself started to get out of control. He said the agency had staffed the inquiry with two part-time field biologists—Helgen and her wetlands partner Mark Gernes—anticipating that they would have to investigate only the handful of sites that had been reported back in 1995. Now that the number of sites had topped two hundred, most of them still unconfirmed and in serious need of field monitoring, he said it was out of the question for the MPCA to try to do it all. He said the agency could keep Helgen and Gernes involved using internal funds, but the "main responsibility" was going to have to be on federal authorities.

"What do you mean, exactly?" I asked.

"We're still planning on doing fieldwork," Anderson said, sounding a faint note of either compromise or resignation, I couldn't tell which. "But we can't become a frog research agency."

I asked Anderson about the accusations being made that the MPCA was running away from a possible fight over agricultural chemicals.

"We're not afraid of the chemical industry," he answered. "But we have to know what we're talking about. If it turns out to be a chemical pollutant then that's what this agency is in business to deal with."

In other words, the MPCA was prepared to deal swiftly with any pollution problem that could be linked to the frog deformities—it was just no longer willing to go out and look for such a problem. This clearly wasn't going to cut it. The only hint of support for the MPCA's abrupt exit from the investigation came in the form of a lead editorial in the *Star Tribune*, which praised the agency for "helping to uncover" the deformities problem and went on to argue that the "extensive research now required to come to clear conclusions is best undertaken by federal laboratories." But the *Star Tribune* failed to offer any advice on just how the feds were supposed to do this, ignoring the fact that it was the MPCA that had the deformed frogs and the study sites. If they didn't play, the game had to start over from the beginning.

While this little soap opera was playing in Minnesota, down in North Carolina it was business as usual. Jim Burkhart told me later that he never even paid attention to the MPCA's threat to get out of the frog investigation. "Resources were really never part of our early discussions with them," Burkhart said. "We just felt that whatever we were going to spend on this would ultimately be determined by what needed to be done."

Burkhart's casual response may have had something to do with the fact that the MPCA was putting a very different face on what it intended to do in private communications with the NIEHS. While the MPCA's plans were in flux in Minnesota, Peder Larson was quietly coming to an arrangement with the NIEHS that was quite different from what he was saying publicly about what his agency was willing to do. On January 29, in a letter to NIEHS director Kenneth Olden, Larson formally requested the assistance of the NIEHS in the frog inquiry. Larson wrote offering a full participation by his agency. "The MPCA is committed to continuing to be a lead agency in investigating this problem within Minnesota," Larson's letter said. "We will continue to provide an active coordinating research role and be directly involved with research planning, sample collection, extensive public, scientific and administrative communications, literature reviews, interpreting sample results, discussion of future research directions and appropriate regulatory response." Was there anything else? Larson's letter concluded by describing the relationship between the two agencies as a "partnering." The NIEHS role, Larson wrote, would be to make a "significant contribution" that would "greatly assist" the Minnesota investigation.

This didn't sound at all like a project being "turned over" to another agency. After a few weeks it evidently sank in that the MPCA couldn't say one thing in public and then take a totally different stance in private. So the agency did another unexpected pirouette and took it all back—with a big boost from then-Minnesota Governor Arne Carlson, a fiscally minded Republican who thought money on frogs would be money well spent. In late February Carlson asked the legislature to authorize $200,000 in new funding for the MPCA to continue frog research in the upcoming season. Carlson released a statement calling the deformities outbreak "one of the most important" problems relating to the environment and human health in the state. The legislature signaled that it would go along with this request, and in the end, so did the MPCA—which ran out of other options once it was clear that it was going to get money for frogs whether they wanted it or not. Peder Larson told the *Star Tribune* that the agency had changed its position and now "looked forward to an expanded role." Putting the cart back in front of the horse in classic bureaucratic fashion,

Larson crowed that the MPCA was in charge once more, just as it was supposed to be. Larson said the extra money would be used to "coordinate" the research efforts of various federal agencies that would be "helping" the MPCA.

Over that same winter I read a slim, wonderfully evocative book by Mike Lannoo called *Okoboji Wetlands*. It's the story of the Iowa Lakeside Laboratory and a naturalist's view of the environs surrounding the lab, which is located on the western shore of Lake Okoboji in northwestern Iowa. This is one of those remote, forgotten corners of the heartland that you can't believe exists until you see it for yourself. Here, where the landscape is quite literally only heaven and earth—is the Corn Belt Riviera, a tony retreat consisting of Lake Okoboji, Spirit Lake, and a number of lesser bodies of water that are all known locally as "Iowa's Great Lakes." Lake Okoboji, a two-mile-long wedge of water shaped roughly like the outline of Great Britain, is rimmed on all sides by lavish cottages, magnificent waterfront homes, and crisp-looking gated communities of town houses and condos. The lab sits atop the wooded slopes bordering Miller's Bay. It's been one of the country's leading field biology stations since 1909.

Okoboji Wetlands is at once a celebration of aquatic ecology and a lament over an ecosystem in decline. Like wetlands everywhere, the ponds and swamps and sloughs of Okoboji have been steadily compromised by the impact of human development. These effects are sometimes dramatic. The act of plowing under a wetland to plant corn is both immediate and permanent in its impact on the species living there. So is the poisoning of a wetland with rotenone to kill "trash" fish species in order to rear game fish in it instead. But these kinds of specific, overt, premeditated alterations of the environment are not necessarily the most important considerations in understanding the more general collapse of wetland habitats, according to Lannoo.

I'd heard Lannoo speak at the Duluth meeting when he moderated one of the discussion groups. I remembered a funny comment he'd made during a debate over how reports of deformities should be confirmed by professional biologists. Lannoo said he was impressed by the reports of deformities from Minnesota and Canada even though he'd never seen a deformed frog in twenty years of amphibian field studies. He agreed with several scientists who spoke up to insist that they needed a lot more data but were worried about what might be involved in getting it. Should volunteers be solicited to survey frogs all over Minnesota and wherever else the deformities turned up? Or was an on-site evaluation by a trained field biologist the only way to get usable information? What would be the impact on the frog habitat of hordes of people wading through swamps and potholes wielding nets and inadvertently stomping on lots of living things that got in the way? The field investigation needed to be carefully thought out, Lannoo had said, lest it damage wetlands in ways that posed a greater threat to the frogs than the deformities did.

"Plus, we don't want to risk having a pack of Cub Scouts out there somewhere watching a bunch of frogs dying of suffocation in a jar," Lannoo said. After a second he added, "Well, actually they'd probably enjoy that. Which may be an even better reason not to use amateurs."

Lannoo, by training a neuroanatomist, teaches human medicine at the Muncie Medical Center, part of Ball State University in Indiana. But because of his long-standing interest in amphibians and extensive field experience, he'd also become the U.S. coordinator of the Declining Amphibian Populations Task Force, an international group of field herpetologists that has been studying the puzzling collapse of frog populations around the world for most of the past decade.

Lannoo returned each summer to Okoboji to teach evolution and wetland ecology. These long summer days were divided between the classroom and daily excursions into the field, where Lannoo and his students would drag long seines through wetlands and then identify and examine whatever they dredged up: snails, insect larvae, salamanders, tadpoles, frogs. What interested me most in *Okoboji Wetlands*, aside from Lannoo's vivid depictions of the rich interplay of life in aquatic systems, was the way he wrote about the insidious nature of gradual changes over long periods of time—years or even centuries—that are not, in terms of the evolution of life on earth, long periods at all. A change in the environment taking place over a thousand years would have about the same significance on an evolutionary time scale as the impact of a comet with earth. Both are effectively instantaneous.

But we don't see it that way because we live in only a sliver of Deep Time. We tend to be captivated by phenomenon that appear to have come upon us suddenly. Our brains, Lannoo wrote, "emphasize novelty." Lannoo noted that the slow but steady winnowing of the pageant of life—the ongoing "biodiversity crisis," as scientists came to call it during the 1990s—had taken hold without much notice largely because few of us can imagine, and none of us can remember, what the world looked like even a hundred years ago. Lannoo argued that one way to bring home the accretion of loss that is the biodiversity crisis is to describe the local environment as it appeared to our grandparents when they were children. Among the "major" species of animals once common in the Okoboji area that are now nowhere to be seen were black bears, buffalo, elk, cougars, otters, and bobcats. You could draw up a similar list of missing animals, not to mention plants, for almost any area in the United States. We take these astounding losses for granted. How easy would it be, then, I wondered, to accept the extinction of a species or two of amphibians? Or more than that.

If you don't care whether there are bears in your neighborhood anymore, would you really lose sleep over the fate of a bunch of frogs? Human beings today tend to live in the sealed, modulated environments of cities and suburbs. Our "environment" is steel and stone and glass and asphalt. We have become forgetful of natural history. Most people nowadays are hard-pressed to identify more than a handful of the bird species that might appear at a backyard feeder; many who would give no thought to navigating a tangle of high-speed freeways would be unable to find their way out of a twenty-acre woods. Not only do we casually accept our ignorance and our losses, we see all of this as the inevitable result of progress. People have come to believe that the vitality of American life is incompatible with the preservation of the natural environment. We've been taught, Lannoo wrote, that environmental concerns are "a drag on the economy. "We pit wildlife against jobs; most often, it's the wildlife that lose against the insistent demands of agriculture and industry and urban sprawl. Lannoo didn't argue that we should turn back the clock a hundred years, only that we needed to see that this approach will not work forever. "We must realize that economic . . . gains that run counter to long-term environmental health cannot be sustained." Our economy, he added, is "nested inside our environment."

Thinking about these observations, it seemed to me that there was something regrettably *trendy* about the investigation of the frog deformities. The grotesque nature of their leg afflictions, plus the attendant worries about a threat to "human health" they might be signaling, gave the outbreak a kind of momentary currency—like the latest killer virus or the newest flavor at Starbuck's. What, after all, do a few limb deformities mean in the context of a biodiversity crisis enveloping the globe? The evidence so far showed that deformed frogs didn't survive to adulthood. But then *most* frogs don't survive to adulthood. They're taken out at successive stages of life by predators, disease, bad weather, and unnatural causes of every description. On average, only about one of every 600 fertilized leopard frog eggs will turn into a sexually mature frog. This is why female leopard frogs lay as many as 5,000 eggs at a time. Frogs are *cheap*.

Maybe the deformities were only another small subtraction at the margin—just a mildly alarming way to do in a few more animals that faced 600-to-1 odds anyway. Obviously, bigger forces were already at work in the world causing wholesale extinctions among plants and animals. And, from the frogs' point of view, if we can even try to fathom what that might be, what's the difference whether you're dead because somebody builds a shopping mall on your home or because your legs don't work right?

Naturally, I wanted to meet Lannoo, whom I thought might be able to shed more light on these ideas. I'd hoped I'd get a chance at Joe Tietge's upcoming meeting in Shenandoah, but when I phoned Lannoo in March of 1997 he told me he wasn't going. Lannoo said he wasn't very happy with the way the frog investigation was taking shape that winter. Tietge had been tinkering with the lineup he'd used in Duluth. He'd belatedly told Lannoo he was welcome to attend the meeting, but that the travel money had all been spent and he'd have to pay his own way.

Lannoo was in no mood for it. He'd just gotten back from a meeting at the National Wildlife Health Center in Madison that he'd found exasperating. I'd heard from Joe Tietge that the Madison meeting was going to take place, but when I asked Kathy Converse if I could attend she had said no. This was to be a workshop for making "strategic plans," she said, and they didn't want any reporters there.

Lannoo indicated that I hadn't missed much, unless I enjoyed watching people make fools of themselves. He said he'd be happy to talk with me about it and mentioned that he would be up my way soon. His in-laws had a summer place in Luck, Wisconsin, and he and his wife had recently bought some property not far from there. Did I know the area? I said I did and that it was only a little over an hour's drive from my house. We agreed to meet there the first week in May. He said he'd be planting trees.

———————————

The eighth of May was a gray, slightly cool day, with a damp breeze blowing smartly out of the south, bending the branches of trees that were still leafing out and riffling the newly green field grasses as I drove up to Lannoo's property. His Toyota 4Runner was parked just off the road. A thinly outlined set of wheel tracks snaked back across a field toward a small grove of trees. I could see what Lannoo liked about this spot. It was a standard forty-acre section, but there was a nice roll to the terrain, which was open on the north and east, and wooded where it rose to a hilltop on west. When I got out of my truck the sound of frogs calling was on the wind in spite of the chill in the air. It was *Pseudacris crucifer*, the spring peeper. Spring peepers are among the earliest frogs to call in the upper Midwest each spring, along with wood frogs and chorus frogs. Each of these species has a distinctive call, but the spring peeper's high-pitched melodic whistle is unusually musical, almost ethereal. Unlike most other frog calls, the spring peeper's never seems to originate from a specific point. The sound of the spring peeper always sounds like it's coming from everywhere at once.

I found Lannoo and his dog, a half-Lab/half-husky named Denali, out on a low ridgeline. Lannoo was carrying a bucket full of needlelike green stalks soaking in water and a shovel. He wore a camouflaged water-fowling parka, which was open and flapped behind him as he dug. He said there was no particular pattern to his planting. He'd gotten a good deal on 1,100 red pine and white spruce saplings. He had the rest soaking in a shallow wetland over in the grove. Lannoo excavated a shovelful or two, dropped in a tree, filled and tamped the dirt back around it, and moved on a few yards to repeat the same procedure. He worked with the reflexive industry of someone determined to plant 1,100 trees by hand in two days' time. While he planted, we talked. I asked him what had gone wrong in Madison.

Lannoo said there had been a sharp disagreement between the field biologists and the toxicologists present at the meeting. In general, he said, the toxicologists wanted to work with model systems in the lab, while the field people lobbied for more intensive studies of the problem where it was actually occurring. There had been an especially pointed debate over the possibility of using *Xenopus laevis* as a laboratory surrogate for native frog species, Lannoo told me. "The problem with toxicologists," Lannoo said as he a levered a hunk of dirt out of the ground, "is that they think a frog is a frog."

This is not the case, he said. Laboratory *Xenopus* might or might not tell you anything that could be usefully extrapolated to wild frogs. Lannoo said the argument hadn't been only about surrogate species, either. Several people at the meeting talked about how data on frog populations were essential—and that this, too, could be gotten without venturing away from the laboratory through mathematical modeling of landscape patterns. When Lannoo had objected vigorously to this concept, one of the particiants pointed out that she could model wild frog habitat using satellite mapping.

"I said that was fine," Lannoo told me. "She's got satellites. You know what I've got that's better?"

"What?"

"Waders."

We made several trips over to the grove to replenish the bucket with saplings, the sound of spring peepers following us everywhere as we trudged along. Lannoo showed me around the rest of the property. Over the hill and down a steep embankment there was a larger wetland hidden away in a low place in the forest. The views from the hilltop itself were breathtaking. "I think we'll build the house about here," he said. He glanced at me sideways as the leaves above us rustled in the wind and added, "Someday."

Lannoo said it was clear to him that the main share of any resources that were going to end up in the frog investigation would be grabbed by the toxicology bunch. He felt the herpetologists were being consulted more as a courtesy than out of any sense that their input was actually needed to solve the problem. It was getting to be an old story, Lannoo said. The ascendancy of molecular biology was reducing interest in the natural histories of whole organisms to the vanishing point. The gene jockeys brought in huge amounts of grant money to their universities, and the resulting financial leverage always fed more research in a similar vein. Even though field herpetology was a relatively inexpensive brand of science, Lannoo said, it was getting harder to fund it all the time. But he didn't see how the frog investigation could progress without a strong input from herpetologists. "There's no substitute for good fieldwork," he said. "We need to establish where these deformities hot spots exist and document what is occurring in them. Dave Hoppe needs funding, not somebody hunched over a lab table."

Lannoo agreed that the frog problem presented a good news/bad news proposition for scientists. The good news was that it would undoubtedly spill over into the more general area of concern and elevate a very real, far-reaching problem—the dramatic global decline of frog populations—to a more visible status. The deformities problem seemed likely to rub off on the population decline issue in a positive way. The bad news was pretty much what I'd surmised that winter. The deformities outbreak was too sexy to resist. Researchers are prone to piling on when a fashionable research vehicle is discovered, Lannoo said.

"Like what?" I asked.

"Well, AIDS is a good example," Lannoo said. "When AIDS was identified as a big problem you suddenly had a lot of people who were doing
research in very peripheral areas jumping in line to get some of the funding. Whatever they had been working on suddenly became AIDS research."

This caught me up short. Lannoo was the second person I'd spoken
with in the space of a few weeks who had compared the deformities outbreak to AIDS. The other was Pat Schoff, from the Natural Resource
Research Institute in Duluth, who'd brought up the AIDS epidemic in an
offhand manner one day when I was visiting him in Duluth. He'd said that
in some respects, the frog deformities were similar to the AIDS epidemic in
that they exposed a certain kind of urgent scientific inquiry to public view.
Schoff said he thought this actually provided a reasonably coherent picture
of the way science "works all the time" when people aren't watching so
intently. I'd asked Schoff if he thought the attention was warranted—
whether he agreed with the proposition that however captivating the
deformities problem was, it was probably warning us of a genuine risk for
human health. He said he did.

"Then why is it that we aren't seeing people from the same regions
developing abnormal limbs?" I asked. This was hardly an original thought.
It was the question everyone had been asking about the deformities. If
there was as actual risk to people, then where were the victims?

Schoff kind of shrugged at the question. Maybe that would happen,
he said. It could depend on the dose and whether frogs were picking up
something that humans realistically would never be exposed to in the
same way. Or it could be that humans were being affected already and we
just didn't realize it. "There's a key question here," Schoff had told me. "In
animals we have to look to see how a population is impacted. In humans,
even one affected individual is a problem. But you don't necessarily get to
the same result. One difficulty is that humans, unlike frogs, are not very
tolerant of abnormal development. I think it's quite possible that if a developing fetus is affected by the same thing causing the deformities in frogs
the result is a spontaneous abortion."

"A miscarriage?" I asked.

"Yeah." Schoff said. "And it could happen early on so that it goes
undetected. A woman misses a period, or it's a little late, and then things
go back to normal without her ever realizing she's pregnant. It happens all
the time. It's hard to quantify and even harder to trace as to the cause. But

I can imagine quite easily that some effect in humans related to the same factor causing the frog deformities could simply get lost in the background noise of unsuccessful pregnancies."

This, of course, is the crux of the problem in human health concerns: What do we define as a problem? How do we measure it? Are we more worried about a dramatic impact that occurs now, or a longer-term insult to our health that spreads out its effects over time? Incremental changes—like the gradual rise in genital abnormalities or the steady decline of sperm counts being attributed to sustained endocrine disruption—ultimately take a real toll in terms of human health. But trends make for vague headlines, as do effects that go unseen. A baby born with three legs would be disconcerting, to say the least. But an abnormally developing fetus that aborts at an early stage may well go unnoticed.

Schoff was not the only scientist in Minnesota who'd been thinking that the effects of chemical contaminants on human health might be largely invisible. Another was Vincent Garry, a professor of toxicology and pathology at the University of Minnesota medical school in Minneapolis. In the spring of 1996, Garry published the astonishing results of a study he'd made of human birth defects in Minnesota.

Garry initially found that in rural Minnesota, especially in the heavily agricultural western part of the state, there was an elevated incidence of birth defects among the general population. Next, he looked at the frequency of birth defects that occurred in the families of licensed pesticide appliers. These numbers were higher still. The abnormal births tended to occur more often in infants who were conceived in the springtime, when pesticides were most heavily used. There was more: The overall birth rate among pesticide appliers was only around one-half of that of the rest of the population in the same areas, a finding that could reflect lower rates of conception or higher rates of early miscarriage. Or both.

Registered pesticides can only be used with a license, and in Minnesota those licenses must be renewed every three years. The state knows who these people are. About 8,000 are commercial pesticide appliers. Another 30,000 or so are farmers. When I phoned Garry in the fall of 1996 and asked him if he thought there might be any connection between his findings and the outbreak of deformities in frogs, he said categorically that it would be pure speculation to think that. Human beings and frogs were different, he said. "Still, it's a temptation to wonder about it, isn't it?" he said.

Lannoo thought pesticides were suspect, too, but seemed much more preoccupied with what the right approach to the problem would be. He thought it was clear after the Madison meeting that the agencies that had stepped forward to work on the problem were going to make it "fit in" with existing research objectives. Maybe that in itself would work out, he said, but he was worried that it put toxicological considerations ahead of all others. "It's all about emphasis," Lannoo grumped. "If they don't want my advice, then they shouldn't ask for it."

Lannoo stopped planting to have lunch. We sat in the shade of our vehicles and talked some more. He offered me half a sandwich and stretched his legs out and fed a bite or two to Denali. The sun had begun to shine faintly through the clouds. It felt a little warmer. Lannoo said he thought a lot about what kind of world he and his wife were making for their son, Pete, who was two. He told me that from an environmental standpoint, "we've already gone off the cliff," but that there were still many options for restoring some kind of natural order to the world.

I'd been hearing this for months as I talked with biologists about the frog problem. Somehow it seemed to keep everyone soldiering on.

"Do you really think the environment is in terrible shape?" I asked Lannoo.

"Sure," he said. "I think in some ways it's too late for us. But I'm still an optimist. The future could be better than we think. It's not too late for Pete."

We chewed our sandwiches in silence. The wind rushed over us, and Denali blinked slowly.

"So," Lannoo said after awhile. "Tell me about Shenandoah."

Joe Tietge's two-day workshop at Shenandoah National Park had taken place in a quaintly aged, wood-framed conference center on the grounds of the Skyland Lodge, which sits high atop the Blue Ridge Mountains near Thornton Gap in Virginia. It was unseasonably cool for mid-April; a number of the herpetologists who had hoped to find locally abundant salamanders in the woods during their breaks were disappointed. The dogwood was blooming at lower elevations, but the air on top of the mountain was bracing. Darkness fell suddenly each evening, and in the black night sky the comet Hale-Bopp rode low in the northwest, its brilliant cloud of a tail pointing straight up and looking so close that it seemed you might reach out and touch it.

In most ways, it was the same meeting Joe Tietge had put on just seven months earlier in Duluth—with the first day again highlighted by gripping field reports from the past season by Dave Hoppe and Martin Ouellet. Hoppe said he continued to puzzle over the higher incidence of deformities that occurred in the more aquatic species, though he cautioned again that all of the affected animals were totally aquatic during development.

But there was now a great deal of discussion about the general quality of the field data, and, in tone at least, the meeting had a very different feel from what Lannoo had described to me of the Madison meeting. Most of the scientists at Shenandoah seemed to agree that there was insufficient field data on the deformities to say conclusively that the phenomenon was a new one that had suddenly appeared in wetlands across the northern United States and southern Canada—although that seemed to be the case. Most people thought the deformities looked highly unusual—there were still many more herpetologists who'd never seen a deformed frog than the other way around—but nobody was certain what the "normal" or background rate of limb abnormalities might be. This number, whatever it was, gained importance over the course of two days of intense discussion. Everyone seemed to recognize the importance of determining that the bulk of the deformities being reported were indeed something new. Not everyone agreed that the data so far answered the question. Ouellet again referred to the extensive scientific literature that had reported on amphibian deformities for decades. He said he'd already collected more than one hundred papers in French that dealt with the subject. But he said that didn't mean that frog deformities were by any means a routine feature of nature. Ouellet agreed that more fieldwork was needed to determine the extent of the problem. But he said that based on his own field surveys— which he was working hard to enlarge—he felt confident that the background rate of deformities in most places was probably only around 1 percent, and very likely less than that. When I asked Hoppe about this, he nodded in agreement.

A lot more of the discussion at Shenandoah concerned population declines among amphibians. One of the more compelling talks was given by David Green, a herpetologist who was curator of the Redpath Museum of Natural History at McGill University in Montreal—and also Martin Ouellet's faculty adviser. Green had been studying toad populations in Canada for years and was impressed with the fundamental instability of amphibian populations on a season-to-season basis. The problem with frogs, said Green, is that they have "very high birth rates and also very high death rates. That means a given population can fluctuate significantly from year to year."

Green explained that there are three types of extinction. Local extinctions involve only a single population of a single species. Then there are extirpations, in which a "metapopulation" covering a region is eliminated. Finally, there is extinction across the entire range of a species. This is final; once all individuals in a species are gone everywhere that the animal is found then the species can never come back.

"And all three of these types of extinction do happen normally," said Green. The point, which he allowed to sink in without any additional amplification, was that it is awfully hard to determine what is going on in a small sample of wild amphibians. Death and deformities are always going to be part of a very large picture. One had to be careful about inferring too much from data that represented only a miniscule snapshot of what might be taking place across a wide area involving many discrete populations of animals.

One person at the meeting felt that things were becoming a lot more clear than that. Gil Veith, the EPA's director of ecological research who'd come up from Research Triangle Park as Joe Tietge's main audience for the conference, told me that everything he heard seemed to indicate that the Duluth team was already on the right track. Veith didn't mention methoprene specifically. But he said it seemed clear that if you considered the basic mechanisms of limb development you could begin to limit your search for causes of the deformities by focusing on agents that could interact with those processes. Vieth said he thought the EPA would be able to "zero in on a small number of plausible causes very quickly." He said he was impressed by how much was already known and that there was a good chance they could come up with some definite answers even faster than usual for such a perplexing environmental problem. It wouldn't happen overnight, he said. But it wouldn't take forever, either. "If we can identify the true hot spots," Veith said, "we have the methodologies to unravel a cause."

"What do you think it might turn out to be?" I asked.

"Don't know," he said. "But certainly there's been a lot of interest generally in the retinoic acid pathway as a target of teratogenic activity. I think we'll definitely want to develop a retinoic acid receptor assay to look at those kinds of agents. I think it's fairly reasonable at this point to at least suspect that a pesticide might be involved. Especially one that is targeted at insect development and that functions as a hormone disruptor."

Again, he didn't say methoprene explicitly.

The next morning before the conference reconvened I had breakfast with Green, Ouellet, and also Stan Sessions—who'd been summoned once again to give his parasite talk. Green amplified on his presentation from the day before. Green told me he'd been studying a single population of toads for ten years. The only way to get an accurate count of the animals was to be there every May at metamorphosis and actually record each animal as it emerged from the water. You catch all the toads you can, he said, marking them with a "toe clipping," a quick, harmless procedure in which the tip of a toe is amputated with a small scissors. You go back, Green continued, and do it over again the next day and the next, noting the animals already marked and marking the new ones, until you stop catching any you haven't seen before.

Amphibian populations, Green said, were always "crashing and rebounding." There were lessons in this for the people working on the deformities problem: "I don't think we've invented this outbreak of deformities. I think it's a real thing. How important it is isn't clear yet. Or how dangerous." Green said he'd be especially worried if some kind of abnormal development that usually happened only infrequently were being amplified. "If we're doing something to alter environmental conditions so that a normal but rare occurrence is being increased to abnormally high levels—well, that would be bad."

Here was a take I hadn't heard before. Green was concerned that something *unnatural* might be causing an increase in the *natural* incidence of frog deformities. Did that mean that such new outbreaks were normal or abnormal?

Green said it would be damned hard to determine. He said it was important to understand that in nature, something is always winning the lottery. Impossible odds can be beaten with impossibly large numbers. "Look," Green said, "the realm of what's 'normal' is much bigger than people think. It's possible that a deformed frog is maybe a one-in-a-million shot. But there are millions of frogs and there are millions of places where it can happen."

This seemed to sit well with Sessions, who was listening quietly. Sessions had also spoken the previous day and his position had hardened remarkably since the Duluth meeting. Sessions now maintained that "the bulk" of the deformities being reported were in fact "natural, normal phenomena." To the extent that deformities seemed to be increasing, Sessions argued that this was almost certainly correlated to an increase in snail abundance. As the first intermediate host in the parasites' life cycle, snails could trigger outbreaks of frog deformities when their own populations exploded. Sessions also hinted that he had come to believe that parasites could cause not only extra legs but *missing* legs as well. He said trematode cysts could be so densely "packed" into the limb bud that no development could take place.

Sessions didn't stop there in his talk. In Duluth he'd surprised people by allowing that parasites might not account for all of the deformities. But now he reversed himself completely. Not only did he think parasites were the one cause, he lobbied for an end to any further inquiry into other possible causes. This hadn't gone over well at all. In Shenandoah, Sessions basically urged everyone to cease and desist in their own research and join the parasite cause. This was presumptuous—and more than a little fanciful, as it was clear that most of the scientists at work on the problem had no interest in studying parasites. Parasites might well cause some deformities, but no one believed they could explain everything or that they would pose any plausible health risk. Why worry about them?

Sessions was undeterred. "What I'm advocating is that we should leave the frogs alone from here on in," he'd said. "We should instead go out and look at snails. Snails are the kingpin in this phenomenon."

It was a strange presentation and it became the talk of the conference. Sessions had literally suggested that no one else was entitled to an opinion contrary to his own—a position that was in no way justified by the experimental findings he'd published in "Sessions and Ruth" years ago. His old colleague, Ken Muneoka from Tulane, was one of several scientists present who took sharp issue with Sessions. "Stan, you've got a very tidy little story," said Muneoka during a discussion period. "But you don't have a shred of evidence to support it."

I didn't think that was quite true—"Sessions and Ruth" was not completely without merit, even if the experiment involved only inert beads rather than actual trematode cysts. But Muneoka was merely making a point of where, in the larger scheme of things, parasites actually belonged. Sessions had slim evidence that parasites might cause extra limbs in amphibians. Although he'd observed and recorded other kinds of deformities at his original sites in California back in the mid-1980s, Sessions had never elicited any of these phenotypes in his laboratory experiments. Based on the fieldwork by the MPCA, David Hoppe, Martin Ouellet, and the researchers in Vermont, multilegged frogs—as intriguing as they were—did not appear to be a significant part of the deformities outbreak. Multiple legs were not among the most common deformities Hoppe had discovered at CWB, which was probably the hottest hot spot in the United States. Instead, skin webbings and anteversions (twisted legs) showed up much more commonly. So did missing limbs or partial limbs. The story was much the same everywhere else. Sessions was proposing that everyone abandon the search for other causes based on a hypothesis that *might* account for only a handful of the deformities that were turning up. Muneoka and others were outraged. "Are parasites a credible hypothesis?" Muneoka said to me during a break. "Sure. But it's really unlikely. The data are very weak."

Muneoka said he was still convinced that the main cause of most of the deformities would turn out to be a "human impact," probably some chemical or chemicals that could mimic retinoic acid or one of the other growth factors or hormones involved in limb initiation and patterning. But he was adamant that Sessions had put the emphasis entirely on the wrong end of the scale of possibilities. Rather than walk away from the problem because a small part of it might not be a problem at all, Muneoka proposed that it would be irresponsible to ignore the possibility of a chemical cause for any of the frog deformities. This was a serious concern, he said. If parasites caused some frog deformities, then that was an ecological novelty that would interest herpetologists and parasitologists. But if chemical contaminants were involved, then the implications for other forms of life, including humans, demanded an intense, well-funded inquiry. In effect, Muneoka felt Sessions had it all backwards. Researchers could afford to ignore parasites, but not other potential explanations for the deformities. "Retinoic acid is a known human teratogen," Muneoka said, tapping his fist silently on the table. "If there are mimics for it out there in the environment we need to know that, whether they're the cause of all these frog deformities or not."

Sessions told me later that he felt Muneoka and others at the meeting had "turned" on him—a claim I found baffling, in that the mocking, antagonistic tone of his talk appeared to invite as much. Didn't he realize he'd provoked such angry responses? Sessions had plainly meant to bait everyone else. For Sessions, it seemed the cause of frog deformities had become an intensely personal matter. His attempt to bully the other meeting participants had gotten what it deserved. A curt dismissal.

Now, at breakfast in the clean, well-lit dining room of the Skyland Lodge, Sessions attempted to moderate his rhetoric. He was smiling and relaxed and listened intently as Ouellet described again a situation in Canada that seemed quite different from what Sessions had seen at Aptos. But he remained insistent that the main investigation into the deformities was, in his view, an utter folly and a waste of time and taxpayers' money. "We're all primed to think in terms of chemical pollutants," Sessions said. "It's understandable. But that doesn't make it true." Sessions paused for a moment, then archly turned Muneoka's assessment around the other way. "Right now," Sessions said, "there's not a shred of evidence for chemicals."

Ouellet, who had been quietly listening to all of this, thought otherwise. His data showed an association between the deformities and the use of agricultural pesticides. It was, in his favorite expression of certainty, "clear-cut." Besides, his experience with multilegged frogs in rural Quebec was the same as Dave Hoppe's down in Minnesota. He didn't see that many extra legs. "Stan," Ouellet said patiently, "even if you're right and parasites cause one hundred percent of the extra hind limbs, that will only explain deformities in less than three percent of my cases."

Several speakers at the Shenandoah meeting hadn't been at the Duluth conference, and a couple of them were plainly stunned at the polarization that had already crept into the deformities investigation—which, after all, was still only in the planning stages for the upcoming season. One of these people was Andrew Blaustein, an ecologist and animal behavior specialist from Oregon State University. Blaustein, a blunt, funny, former New Yorker, was known for his head-on style. He was a quick student of scientific enigmas and the politics that can accompany them. He had become, over the last half-dozen years, a leading figure in the study of amphibian declines because of his work on the effects of increasing ultraviolet radiation on frogs living at high altitudes in the western United States. "I'm sorry to tell you all that, in spite of what Rush Limbaugh says, the levels of ultraviolet radiation reaching the earth are increasing," Blaustein told the group. Elevated ultraviolet penetration of the atmosphere—in particular the segment of the ultraviolet spectrum called UV-B—was already causing problems in amphibian populations among certain species. Blaustein said the effects were most clearly seen in species that laid their eggs in open shallow water at higher elevations.

Blaustein had gotten onto the problem when he began having trouble finding formerly common mountain species such as *Rana cascadae*, the

Cascades frog, which lives at elevations of between 3,000 and 9,000 feet in the mountains of Oregon and Washington. Blaustein had proven that the Cascades frog was one of several species of amphibians that has an inefficient cellular mechanism for repairing damage to its DNA caused by exposure to UV-B. He said that in some populations, egg mortality in *Rana cascadae* had begun to approach 90 percent. Frogs are cheap, but not that cheap. Blaustein said there was good historical data indicating that egg mortality in a healthy population should be more on the order of 10 percent. His suspicions intensified, he said, when he discovered that if he took egg masses back to his lab they would hatch normally. He'd since determined the mechanism of DNA repair that appeared to be the critical link between UV-B and population declines, and he'd also demonstrated that UV-B could cause certain kinds of developmental abnormalities, although he'd never seen limb deformities among those effects. Blaustein's findings were widely known and a little controversial—the exact amount of UV-B increases taking place in North America is somewhat conjectural—but he was a compelling speaker and his concerns were more than apparent in his personal style. He was dressed in black from head to toe, with long sleeves, and had arrived in the gray light of dawn already wearing dark sunglasses.

Blaustein didn't see any need for the deformities investigators to divide into warring camps. Speaking during one of the discussion periods, Blaustein said he'd heard enough to believe that more than one approach to the problem was needed. He said he didn't want to look at any more "gross frog pictures" because they made him sick. I got the sense he felt the same way about the bickering going on between Stan Sessions and some of the other scientists. "I think parasites and chemicals are both plausible causes," Blaustein said sharply. "They ought to be considered equally." Unlike a lot of the other people in the room, Blaustein tended to stop talking once he'd made a point. This one seemed reasonable enough, and I saw a number of people nodding their heads.

"Just a minute," someone said. It was David Gardiner, a developmental biologist from California who'd been the last speaker on the agenda the first day of the conference. Gardiner, an imposing figure, didn't look quite like any of the other scientists in the room. He stood about six-four, and although he was in his late forties his close-cropped blond hair and lean physique made him look more like he'd just stepped off a surfboard than from behind a microscrope. Gardiner was new to the frog investigation, but there wasn't much doubt as to where he stood, at least hypothesis-wise. Gardiner believed certain, specific biochemical pathways either made limbs or didn't make limbs, and that a deformed frog told you unambiguously that something had messed with those pathways. Gardiner's talk had paralleled Muneoka's, which in turn was much the same discussion of limb pattern formation he'd given in Duluth. Gardiner emphasized that intercalation—the means by which cells in the limb field always attempt to maintain their positional identity by communicating among themselves—was in fact a basic organizing mechanism. It occurred throughout embryonic development as the fertilized egg proceeds through critical phases of orientation. The chemical signals that regulated gene expression in developing limbs weren't simply an anomalous event that occurred only when it was time to build legs, but were better seen as a part of the river of effects that commenced at conception and flowed onward as the embryo grew and developed specialized tissues and organ systems. Gardiner said he was convinced that retinoids were the only cause that had been proposed for the frog deformities that had previously been demonstrated to produce all of the phenotypes being observed in the field. That didn't mean the cause was retinoids, he said. They just ought to be high on the list of things to check out. Gardiner suggested that other disruptions of different biochemical signals that occurred "upstream" in the cycle of development could also result in limb defects at later stages. Like damming or undamming a river, changes would occur downstream. The signaling and gene expressions that took place throughout development cascaded forward onto one another; some factor that acted as a "precursor" to limb initiation could also be involved.

It was a difficult presentation, not least for coming at the end of a long day. But it packed a wallop—Hoppe told me afterward he thought it was basically a "smoking gun." Gardiner told me later he hadn't really known what he was supposed to talk about, but that he had come to the meeting determined to reclaim certain ideas he thought Sessions had coopted to support the parasite hypothesis. This was not an entirely a dispassionate, scientific position. It was personal. Like Muneoka, Gardiner had a history with Sessions that went back to their days as postdoctoral researchers in Susan Bryant's limb lab at the Developmental Biology Center at the University of California Irvine. Unlike Muneoka and Sessions, Gardiner had stayed on at Irvine. He had his reasons. One was that he preferred the independence of being affiliated with a major research lab but without the mustiness of a faculty position. Another was that he was married to Sue Bryant. Gardiner had coauthored a number of the seminal papers on limb development that Sessions was now turning into a foundation of support for parasites.

Gardiner didn't know anything about the frog deformities that were being reported now, but he knew all about "Sessions and Ruth." Ironically, it had been Gardiner who'd suggested the bead experiment when Sessions was puzzling over how parasites might be involved in the deformities he'd found near Aptos. Years later, Gardiner was now dismayed that Sessions apparently had set out to forestall any other lines of inquiry. "I wanted to let people know that Stan hadn't thought up intercalation on his own," Gardiner told me. "I wanted them to know that he was not the expert."

As the light over the Blue Ridge began to fail, Gardiner stood up to object to Blaustein's contention that parasites and chemicals be given equal treatment. Gardiner explained that he assumed it wouldn't be possible to adequately fund investigations into both causes. In a world of finite resources, he said, you'd better spend money on the stuff you're really worried about. It only made sense, Gardiner maintained, to target research into causes for the deformities that might give rise to human health concerns. This was the point Muneoka had made to me earlier, but Gardiner finished it up with a flourish. He said he thought it was important to go back to seeing the deformities outbreak as a kind of environmental distress signal. Whatever the cause, it seemed to say that the world was changing in ways we would have to struggle to keep up with.

"We're living in the middle of a big experiment," Gardiner said. "We put all kinds of newly synthesized compounds out into our environment and then we wait to see if we get a result. Well, this seems to be one. These deformities may be a result of our grand experiment with chemicals. And now that we've got this result we have to figure out what it means."

TWO

Famous Last Days of the Golden Toad

10 Uncertainties

SPRING. SUMMER. FALL. THE SEASONS OF THE FROG IN NORTHERN North America came again in 1997, and with the return of the frogs the deformities investigation began anew. Stan Sessions's call for an end to the inquiry went unheeded. It was the same all over. In the lab and in the field, in the East and the West and everywhere in between, a swelling army of biologists and toxicologists joined the chase.

Everyone seemed to find something that looked tantalizingly like a result, although the meanings of the findings were more elusive than anyone supposed they would be.

The news from the field remained mystifying. A patchwork of deformities reports blossomed in a band across the continent's upper latitudes. For the first time an effort was under way to bring coherence to the reporting process. At the Shenandoah meeting in April 1997, the United States Geological Survey announced it was setting up a Web site it hoped would function as the central repository of deformities data. The National Reporting Center for Amphibian Malformations (NARCAM) went online officially in June of 1997 and a torrent of deformities reports immediately ensued. (The word *malformations* had been decided on as a more precise descriptive than *deformities*, since it implied a developmental origin of the abnormal morphologies. But despite a general agreement that this was from now on the preferred term, everyone continued to call them deformities.)

In its first year, NARCAM received some 560 reports of abnormal amphibians. Deformed frogs, toads, and salamanders were discovered in forty states and three Canadian provinces. Quite a few of these sightings involved only one animal at one site. Their significance was questionable—although subsequent readings of the exhaustive literature on frog deformities showed that this was fairly typical of the historical record. But there were plenty of reports of deformities in numbers too high to ignore. Most concerned frogs and most of them, too, were clustered around the Great Lakes region. The highest frequencies occurred in Minnesota, Wisconsin, Vermont, New York, Missouri, and Michigan. Later Massachusetts, Connecticut, and New Hampshire joined the list of places with high rates of abnormalities, as did several states in the far west—Washington, Oregon, and California.

I was preoccupied with another kind of data that showed up on NARCAM: negative findings. At Shenandoah, the biologist in charge of the project, Doug Johnson, had said NARCAM would not be restricted just to reports of deformities. "Our intent is that NARCAM will create a continual survey of herpetologists," Johnson said. "We want to include information from field researchers who are handling large numbers of amphibians and are not seeing deformities."

This was a tacit admission of a question that many researchers privately conceded was still an unresolved concern. The whole deformed-frog investigation had been undertaken largely on the strength of personal testimonials from a handful of field biologists—scientists such as Dave Hoppe and Bob McKinnell—who'd examined thousands of animals in the field over periods of many years without ever encountering an outbreak of deformities. Nobody doubted the accuracy of their observations; the collective experience of attentive, seasoned scientists was more than relevant—it was about all there was to go on. But the world is large and the ranks of whole-organism biology were dwindling. Could the impressions of the relatively few biologists working on frogs in the field be relied on as a comprehensive gauge of what occurred in nature? Bluntly, could the deformities be—as Stan Sessions had put it at Shenandoah—a "natural, normal phenomenon" that simply hadn't been widely noted before?

Isolated outbreaks of limb deformities in frogs were a natural fact of life—that wasn't in question. Few biologists had ever observed such an outbreak, but everyone had read about them. Besides "Sessions and Ruth," the most widely talked-about paper on amphibian abnormalities that winter and spring was an article published in 1974 in the *Journal of Herpetology* by biologist Leigh Van Valen of the University of Chicago. "Van Valen," as the paper was called, was a rumination on the ecological and genetic causes of frog deformities involving extra legs—a condition known technically as polymelia—and the possible role such abnormalities might play in evolution. But most readers were more impressed by the author's extensive citations of prior outbreaks of deformities. One line in the article was a particularly chilling reminder that nature occasionally mixes in a curve ball: "Isolated cases of polymelous frogs, toads, and more rarely salamanders are reported at frequent intervals, often by people who are unaware of the numerous previous reports," Van Valen wrote.

This was an eerie echo of the cautionary language used by another group of researchers who years earlier reported in the herpetology journal *Copeia* on their 1958 observation of extra legs in about 20 percent of a population of Pacific tree frogs—the same species later studied by Stan Sessions at Aptos in California. The deformed frogs were discovered in a small mountain pond near Polson, Montana, and turned up there again in similar numbers in 1960 and 1961. The scientists were at a loss as to the cause of the abnormalities—although they noted that the pond was contaminated by watering cattle and, curiously, was also found to be unusually radioactive. Nor could they explain why other frog and salamander species that lived in the same area of the pond appeared normal. But their main puzzlement seemed to be a nagging suspicion that the deformities, while rare, were a biological feature of nature without special significance. "The occasional appearance in a collection of amphibians of a specimen possessing more or less than the usual number of limbs does not ordinarily arouse enough interest to warrant a descriptive note," the scientists dryly noted.

"In this case, however," they continued, "the relatively high incidence of extra hind limbs observed in a small population . . . justifies a brief report on the basis that other herpetologists might care to join in speculating on the causal factors involved."

So there it was. Deformities happened all the time and, so far as anyone knew, for all time as well.

Even so, limb abnormalities in frogs dodged precise description as a single phenomenon. Variety seemed to be the hallmark of the current outbreak. Missing limbs and extra limbs—sometimes seen together, usually not. Skin fusions. Anteversions. Truncations. Reductions. Stubs. Spindles. Corkscrews. Tapering, tentacle-like stuctures that could only approximately be described as legs. The historical record was one of singularities: one type of deformity, in one species, and usually for only one season. The current outbreaks were widely distributed and were expressed every which way in the afflicted frogs. And they lingered from one year to the next. Everything that was now being seen had been seen before, somewhere, sometime—but not in such consistently high numbers nor in such an apparent jumble of phenotypes. The Montana group thought limb abnormalities in "a specimen" would not be worth reporting, but finding more than that had caused them to pause and take notes. Decades afterward at Shenandoah the argument turned on the "baseline" rate of deformities. What everyone wanted to know was how often abnormal frogs show up in normal populations. But there was another kind of baseline to consider. How often did abnormal populations show up in the world?

Apparently, not very often—even now. Despite the rising tide of deformities across a vast part of the continent, it was soon obvious that they remained uncommon. As the deformities reports streamed into NARCAM via e-mail and telephone during the center's first twelve months they were outnumbered by reports of no deformities—more than 620 negative reports in all. Since it seemed likely that most deformities would be reported while most normal frogs would simply be ignored—and that both kinds of numbers represented only a vanishingly small fraction of all frogs—the frequency of normal frog populations was no doubt understated to a large extent. Normal was still the norm.

I thought about what all of this could mean many times during the summer of 1997 as I traveled across the country to follow the season's research. It was confusing. The findings seemed to add up one minute and go to mush the next. The size of what was not known about frog ecology seemed to grow exponentially—questions about the dynamics of frog populations were hotly debated but often dissolved unanswered. Would high rates of deformities cause wholesale population crashes? Some people thought yes, some thought no. Most thought it was impossible to say. Were different types of malformations caused by different agents? Nobody knew. At what stage could a tadpole lose a limb and regenerate a complete replacement? A partial one? Those experiments had never been done. Was it even important to know? Could attempted predation cause missing limbs in a significant portion of a frog population? The great majority of biologists and herpetologists I talked to said no. No way. But the question refused to go away.

Sometimes in the field you catch a glimpse of a frog out of the corner of your eye as it darts from an opening and disappears into the tall grass. It happens so fast that you register the movement in the instant before you realize it was a frog. Presently, you may wonder whether what you saw even was a frog. By then, of course, the frog, if that's what it was, is long gone. Only the image remains, like so much green and brown ectoplasm. The outbreaks of deformities across North America took on the same vaporous appearance the longer you looked at them. Was something really going on out there or not? Limb deformities are normal. They are natural. They just didn't appear to be common. Nor did they occur everywhere, even now. Whatever was happening didn't fit the historical pattern—yet it offered no distinct outline of its own. The deformities outbreaks had locations, but in a larger sense they seemed only to exist somewhere between the commonplace and the unknown, a place that was not fixed, but motile and elusive.

No wonder the skepticism persisted, even among the scientists working ardently on the problem. One time when I was visiting with Joe Tietge at the EPA in Duluth he confessed to his own doubts.

We were walking from his office down to one of the labs in the EPA building, a sprawling brick structure of three stories and a basement. Built in the squat industrial fashion of the late 1960s, it has a bunkerlike quality to it—notably in its tall, narrow windows that fail utterly to take advantage of what could have been breathtaking views of Lake Superior only a few hundred yards away. In many visits there I never did get a comfortable sense of where things were inside. Tietge and his colleagues often led me on circuitous excursions down long hallways and through mazes of doors and stairways that made me feel as though I was passing through the opening sequence of *Get Smart*. Tietge and I were going to look at an experiment that was underway—the EPA had by then committed about $500,000 to investigate the deformities—when suddenly he stopped at the head of a stairwell as another door swung heavily shut behind us with a metallic clunk.

"You know what I wish?" he said, changing the subject abruptly.

We'd been talking about the hope that with so many people working on the problem, there might be some answers just around the corner.

"What?" I said.

"I just wish we knew what we were dealing with," he answered. He blinked once or twice but remained pretty much poker-faced.

"You mean what the cause is," I said.

He considered that for a moment. Then he shook his head.

"No," he said. "I mean whether we're working on something that is a real problem."

The Poultney River winds around a thumblike protrusion of Vermont that juts into Upstate New York, forming part of the border between the two states just below Lake Champlain. Not far from the little town of West Haven, on the Vermont side, the river passes beneath the sheer granite face of the foothills of the Green Mountains. Along this stretch the course of the river is followed by a narrow, rutted gravel road that separates it from a large, swampy wetland and, farther beyond, farmlands that you can't see but from which the sound of cows lowing can be heard. In late July of 1997—about ten months after deformed frogs had been discovered there and at several other sites in Vermont on the heels of the Duluth meeting—the Vermont Agency for Natural Resources (ANR) sent field investigators back for another look. The team, led by Kathy Converse of the National Wildlife Health Center in Madison, Wisconsin, and Laura Eaton-Poole from the U.S. Fish and Wildlife Service office in Concord, New Hampshire, was one of several dispatched that day to begin a re-survey of a number of sites that had produced abnormal frogs the year before. Over the next several days, the Vermont researchers hoped to collect leopard frogs at between forty and fifty different sites across the large area of the state. Television reporter Chris Bury and crew from the ABC news program *Nightline*, which had broadcast a report about the deformities outbreak in Minnesota a month earlier and was now following the expanding story, were in Vermont and went along with the Converse team. So did I.

The deformities outbreak in Vermont sounded several familiar chords. All of the affected sites had been discovered in the 8,000-square-mile watershed of Lake Champlain, a heavily agricultural area that experienced unusually high water levels during a very wet summer in 1996.

At Shenandoah, biologist Rick Levey, who'd responded to the initial reports and was now heading up the Vermont effort for the ANR, said he believed frogs had bred that year at many locations that normally wouldn't hold enough water for amphibian reproduction, but that wherever they came from, the animals would likely have been exposed to runoffs from the more than 3,000 farms in the area, two-thirds of which were dairy operations. In checking through the state's wildlife records, Levey said they'd found only five previous reports of frogs with abnormal limbs in the past eighty years.

Levey picked the Poultney River site as the designated media showcase because it had shown a high rate of deformities the previous fall and he assumed—correctly—that ABC and I would be interested in seeing abnormal animals.

We got there late in the morning, a caravan of cars that jounced along until we came to a place where the river turned away sharply to the south. Dust from our motorcade settled on the leaves of the trees on either side of the road as we got out. The day was perfect, sunny and pleasant. A few high clouds lolled in the pale blue sky.

It wasn't immediately clear where to find the frogs here. Or, rather, where it would be feasible to catch them. The river proper ran fast and straight between steep-sided banks about thirty-five yards apart. It was pea green, with dark eddies that pocked the surface. It didn't look the least bit froggy. The riverbank was more promising, but high—a rolled hill ten or twelve feet above the shoreline in many places—and tangled with low bushes and scrubby trees. A shorter fringe of grass lay flattened and brown atop a low berm that ran along the gravel road. The prime frog habitat, and certainly the place where frogs would have bred earlier in the spring, was in the swamp, which was contained by another knee-high earthen rim running by the other side of the road. The wetland itself was a tight maze of small trees and deadfalls and, deeper in, cattails higher than our heads.

Eaton-Poole, an amiable young woman in a starched khaki uniform, scrambled into the brush along the riverbank, skidded down to the water on her backside and gave a husky shout that there were a lot of frogs down there but that it would be very slow going after them. The camera crew rolled tape as she slogged along the river's edge. Converse and I took a pair of nets and scuffed down the road a short distance, where we were surprised to see many frogs sitting rather jauntily right out in the open, in the middle of the right of way we'd passed over just minutes before. There seemed to be hundreds of them. As we approached and they fled to either side we realized that the roadbed, with its raised borders, was a kind of natural frog trap. The extra jump or two the animals needed to negotiate the sandy berm and escape into the weeds allowed us just enough time to get a net down in front of them.

All of the frogs were good old *Rana pipiens*. We caught them easily, and although we didn't stop to examine each one as it was retrieved from the net, it was obvious that a great many of them had hind limb problems—mainly missing or shortened legs. In spite of their handicaps, the frogs were quite agile and strong. They jumped well and kicked lustily when we held them. I found it impossible to detect any difference at all in the way the abnormal frogs moved as they tried to get away, although I supposed that a more capable predator than a man with a net would have a different experience and that none of these frogs were likely to live very long in the wild. Converse and I made several passes up and down a 200-yard stretch of the road, each time finding frogs that had moved back out to where we had chased them from before. After about an hour and a half we stopped, and Converse and Eaton-Poole began a methodical field examination by the side of the road.

It was a tableau I saw repeated many times in many places that summer: Biologists sitting in the shade of a vehicle in the afternoon and taking up frogs one by one from a cooler or a bucket, looking them over front to back, top to bottom, as someone took notes.

Picked up carelessly, a healthy leopard frog has little trouble extricating itself from your grip. Adults are strong and quick to gain leverage with their powerful hind legs. But even with the much smaller young of the year you have to hold them right. Typically, the frog is grasped in a loosely closed fist, with the hind legs extended in the palm of the hand. This leaves the forward end exposed while rendering the frog largely immobile so it can be turned and examined. The legs, of course, are closely inspected. The forelimbs are almost always normal and require only a cursory side-to-side comparison and toe count. With leopard frogs you can sometimes guess that they have recently metamorphosed by the freshness of the bow tie–shaped skin closures across the chest where the front legs emerged. Next, the frog is usually transferred from one hand to the other, with the head and upper torso now held so the hind legs are exposed for a thorough going-over. The legs are stretched out and matched to each other. Are they the same length? the same

girth? Are they properly segmented and articulated? Are the bones complete and straight? Is the skin all right—smooth and tight against the
underlying tissue and properly pigmented on all sides? Do the legs flex
normally? With its legs fully extended a frog's toes bunch together and
point, like a ballet dancer's. But if you push gently on the bottom of the
foot, so that the ankle moves upward toward the knee, the toes splay
out against their webbings and can be counted and measured and
checked for abnormalities.

If you know what you're doing, a complete external examination
might occupy fifteen seconds—or however long it takes for the person
recording the data to write down what you are dictating. With severely
deformed animals it takes longer to analyze and categorize the abnormalities. The frogs do not seem to enjoy the process. When you first pick up a
frog it feels cool and slick, but before you're done it will become sticky
and warmer. Frogs are ectotherms—their body temperature is always
close to the temperature of their surroundings. They're generally tolerant
of temperature extremes in northern climates, but I don't think I ever
held a frog for a long time without feeling as though it wanted me to let it
go. The exact anatomical order in which different biologists examined
frogs varied. But I can't remember watching anyone who didn't finish up
with the mouth and eyes, and, just before releasing, looking the animal
squarely in the face. Watching Kathy Converse work through our collection at the Poultney River, I noticed that she sometimes said something to
the frogs as well. "All right now," she'd murmur, "you're all done." That
sort of thing.

Converse worked as a nurse before going back to school for her Ph.D.
in biology, and I noticed that she was unusually gentle with the animals in
the field. Sometimes she'd open her hand or set a frog down on her leg
and it wouldn't go anywhere. I'd asked Levey to let me travel with her
team regardless of where they went because Converse had already worked
on frogs from Minnesota and because she planned to take about five dozen
frogs from several Vermont sites back to Madison with her, and I was
going to go with them.

Levey's hunch that the Poultney River site would show a significant
level of deformities proved right. In a smaller sampling earlier in the summer, nine out of fifty frogs collected were deformed. We seemed to have
gotten there right at the peak of the frogs dispersal following metamorphosis. All of the frogs we caught were juveniles ranging in size from 16 millimeters to slightly over 40 millimeters in snout-to-vent length—just right

for young of the year, maybe a tad smaller than their Minnesota cousins would be at the same stage. But size wasn't their problem.

In all, we caught a 121 frogs that day. Only 66 were normal. The rest—almost 46 percent of the total—were deformed, a genuinely shocking frequency. Bury asked Converse on camera what she made of such a high ratio of deformities. "Well, I guess it means we should keep looking at what is happening in this area," she answered. Bury asked her how long she thought it might be before the research effort began to pay off. Converse indicated that, in the near term at least, the best hope was that they could begin to rule out some of the factors being proposed.

"This is the first summer of really large-scale surveying of what is going on," she said. "Hopefully, by the end of the season we'll begin to have some concrete ideas about what isn't the cause. That's usually the way it works."

I thought this was an odd choice of words. If there was anything "usual" about the deformities problem, it wasn't obvious. Presently, however, Converse pulled me aside and showed me something that actually did seem to exclude one much debated cause in at least some of the missing legs: predation.

All of the abnormalities at the Poultney River did, in fact, consist of missing or shortened hind limbs. One unusual specimen, a frog with a shortened "calf" segment in an otherwise normal leg, seemed to be missing the bone in that part of the leg. The normally fused bone called the tibula/fibula, which ordinarily runs from the knee to the ankle, was simply not there. The bottom half of the leg swung limply from the knee. This and the other more typical deformities mostly occurred on just one side, with the limb opposite the abnormality appearing perfectly normal. That wasn't entirely certain, since a remaining leg couldn't be compared to one that was absent. These types of deformities closely paralleled most of the reports from Minnesota as well as Canada. Converse said she was mildly surprised not to see any of the skin webbings or "cutaneous fusions" that were also often found in Minnesota.

But she had found something else she believed was highly significant: an unusual disruption of the pattern in the skin pigment on some of the truncated legs.

Like zebras, which all have stripes but in a different pattern on each animal, leopard frogs share a common color scheme while retaining individual characteristics. The animals are white on the bottom, and green or brown on the top or *dorsal* side, as biologists call it. Against this dark dorsal background, black spots are distributed. Each animal has its own conformation of spots that is individualized and distinct—like a fingerprint—but in all of them the spots merge and elongate into bands or stripes on the dorsal side of the hind legs at right angles to the limbs. These dark lines are quite visible as the frog is viewed from above. When the animal is at rest, with its legs pulled up together so that the calf lies immediately behind the thigh, the bars align precisely into continuous bands across the upper and lower leg segments.

What Converse noticed were several instances in which legs that were truncated had a broken pattern on the remaining portion of the upper leg. Instead of neat, orderly bands of darker pigment like those on the normal limb, the skin was mottled and the stripes irregular or absent on what was left of the truncated limb—and not just immediately adjacent to the tip but continuing on some distance nearer the body. Converse thought this was strong evidence that the leg had not developed properly. This kind of skin disruption, she said, did not appear to be consistent with amputation by a predator. Why, she asked me, would the skin pattern be disrupted above the bite? "Look at this," she said, holding one up for me to see. The frog was missing the bottom half of one leg. The dorsal skin on the remnant was a disorganized swirl of varying pigments. "If this were the result of trauma you'd expect the skin pattern to be normal in the part that wasn't bitten off," Converse said.

This seemed to reinforce the findings that Converse and her colleagues back in Madison had made the previous season in their workups on frogs from Minnesota. Of the approximately seventy animals they'd seen, all were missing legs or parts of legs, and in none of the cases did they find any direct evidence of trauma. In some frogs they did find other skeletal abnormalities that seemed to argue for a developmental problem, but it was the legs themselves, or their absence, that was striking. "It's not like these legs developed normally and something happened," Converse had told me earlier. "It's like they were never there in the first place."

But Converse also thought it was too early to conclude anything about what was going on. She felt Stan Sessions was alienating everyone with his insistence on natural causes for all the deformities and his attempts to discourage research into other areas. "It's the way he presents things," she said. "He seems to want to close the doors to any other line of inquiry." Converse was meanwhile worried about the limits of the ongoing investigations. Little was known, she said, about what was happening among the vast majority of amphibian populations that weren't being monitored and never would be. "Minnesota and places like it have lots and lots of wetlands," she said. "If you've got a big marsh surrounded by cattails, who knows what's going on with the frogs that live in it?"

It was hard enough to understand what was happening to the frogs that you did see, Converse said. A metamorphic frog with a deformed leg had acquired the condition weeks or even months earlier. Routine necropsies and pathology tests performed long after the fact might not be especially useful in determining what had gone wrong during an early stage of development. "In Madison we receive animals all the time that are actually dying right in our hands and we can still have a hard time figuring out what is wrong with them," Converse told me. "With these frogs you're trying to trace a problem back in time."

When we left the Poultney River the sun was still high in the sky. The afternoon had warmed considerably. I rolled down the window and looked out at the surface of the river glittering behind the tress. As we slowly moved off down the road, small frogs scattered in all directions. It was hard to believe. You couldn't see it from only a few feet away, but almost half of them were not the way they were supposed to be.

The Vermont Agency for Natural Resource's three-day survey of wetlands in the end included data from nearly fifty sites in fourteen counties. Only nineteen sites yielded large enough samples for statistical relevance, but seventeen of those places yielded deformed frogs. The frequency ranged from 2 percent to as high as the 46 percent seen at the Poultney River. The seemingly haphazard distribution of outbreaks, and their variable severity, was puzzling and altogether like the pattern that seemed to hold in Minnesota. Why were there deformities here and over there—but not in all the other wetlands in between? Before I left Vermont I paid a visit to someone I thought probably saw more leopard frogs than anybody else in that part of the world, an entrepreneur named Jim Mumley.

Mumley, a shy and gracious middle-aged man, lived in Alburg, a small town on Lake Champlain right below the Canadian border. It's good frog country thereabouts, and frogs were Mumley's business: For thirteen years, he'd owned and operated the J.M. Hazen Frog Company, which supplied leopard frogs for scientific uses all over the country. I'd spoken with Mumley on the phone during the winter and he told me he shipped around 40,000 leopard frogs a year. He recalled that in the past he might see an occasional animal with a missing leg, but that the number seemed to be increasing of late.

I'd asked him on the phone if he thought something unusual was going on. He allowed that it was possible and had been wondering how it might affect business.

"I think there is something to it," he said. "I really do. I just don't know what to make of it. It hasn't had an impact on us yet, but I expect in a year or two it might."

Mumley ran his business out of the basement of his house. He'd invited me to stop by, and when I did, he was politely reserved at first. We drank some lemonade and sat in lawn chairs under a large tree in the front yard of his shaded suburban home. I told him about the frogs at the Poultney River and also about what was being found in Minnesota.

"I don't think we've got anything quite like any of that around here," he said after awhile. "C'mon in and have a look."

I followed Mumley inside and we went downstairs. The J. M. Hazen Company was a modest, one-man operation. Mumley waved a hand at a small workbench that served as his shipping center—there were packing materials and address labels lying about—and took me into a back room that held four large, galvanized stock tanks, each about thirty inches high, three feet across, and maybe twelve feet in length. The air was damp and there was an earthy smell mixed in with the usual sort of basement mustiness. Mumley flicked on a light overhead and I heard a splashy commotion in the tanks. Peering into one I saw the bottom was covered with juvenile leopard frogs in a half an inch or so of water. When I straightened up they spooked again and surged toward the other end of the tank in a jumbled, pulsing mass.

"Wow," I said. "That's a lot of frogs. What do you feed them?"

"Nothing," he answered.

Mumley said he didn't keep the frogs very long and that they could go quite a bit longer than that without food. He said he bought them from local pickers who worked around the area. So far as he could tell, leopard frogs were still plentiful in that part of Vermont. But he indicated that he was a little nervous about me reporting how many he took each year. I'd already quoted him on the topic in an earlier story in the *Washington Post* and he was uneasy. "I just don't want anybody to complain," he said.

I told him I doubted that would be a problem and that I was impressed that he could do all this by himself. I couldn't stop looking at the frogs. "Do you pack and ship them all personally?" I asked.

"Oh yes," Mumley said. "I send them quite a few places."

And were they typically juveniles? He said they were. Sometimes he got requests for adults, though more often special orders were for the smallest animals he could find. "Those are from people who are feeding them to their snakes," he said with a slight grimace.

I told Mumley that I was pretty sure that in a single year he must handle as many leopard frogs as most herpetologists would examine in several decades of fieldwork.

"Hmm," he said. We looked at the frogs rocking rhythmically against each other in one of the tanks.

"What I'm curious about," I said, "is just how many animals you see each year that have abnormal legs."

"Oh. Quite a few, I'd say," he answered.

"Like how many?" I said.

"Well, I suppose forty or fifty. Something like that."

"Out of forty thousand?"

"Yeah," he said. "Sometimes a little more or a little less."

I did the math.

"That's just about one tenth of a percent," I said. "That seems to be well below what scientists think is about the normal rate."

"Well, then, I'd say that's right," he said.

———————

A few days later, I was in Madison, Wisconsin, at the National Wildlife Health Center, which is located on a campuslike enclave in the western suburbs. Among the center's primary functions is to act as a kind of coroner's office for wildlife—a morgue. The deformed frogs caught by the ANR in Vermont, some of which arrived the same day I did, weren't dead, of course, but they were doomed. It was understood that because of their handicaps they soon enough would succumb to either predation or starvation in the wild and thinking about that as the animals were brought into the necropsy room in their shipping boxes it seemed to me that what was about to happen to them was only technically jumping the gun. I watched things being set up for a couple of minutes through a thick, soundproof glass window that looked into the main examining bay. Then I went into a changing area, donned a lab gown and rubber boots, stepped into a sanitizing footbath, and passed through an airlock into the inner chamber.

The room was large, with cinder-block walls and a line of stainless-steel examining tables running down the middle. On one was the large brown carcass of a golden eagle, which appeared to have been shot and was awaiting determination as to the exact cause of death. The air was cool and had a rather pleasant antiseptic tang to it. Two tables away, a slight, intense-looking woman had begun working on the frogs. She introduced herself to me in a clipped monotone as Carol Meteyer, a staff pathologist. She smiled but didn't look up.

"Here we go," she said.

Meteyer had already euthanized a couple of the small leopard frogs and now removed one from a jar that contained a lethal concentration of halozine, an anesthetizing gas. Meteyer had to feel around a bit for the animal and for the forceps she used to move it owing to the 2X-magnification eyepiece she wore, which restricted her field of vision to a narrow tunnel. A small microphone dangled on its wire just above her head.

The animal was limp. It was missing a hind leg. Meteyer put it into a large glass dish, the bottom of which was lined with paraffin so the specimen could be pinned into position as needed. Meteyer depressed a pedal with her left foot to activate the dictaphone and began talking as she turned the frog over and around in the dish. Meteyer described the animal's external features and overall condition, which was normal but slightly emaciated. She carefully measured and recorded dimensions of the limbs, including each segment and extremity, down to the toes. She described the skin, including the skin on the hind flank where the missing leg should have been, as "unremarkable." Hunching over, she peered intently into the frog's eyes. Her own eyes, I noticed behind the magnifying glasses, were electric blue.

"Unremarkable," she repeated.

Meteyer opened the animal's mouth, examined its jaw and tongue and larynx.

"All unremarkable," she said.

Meteyer sliced open the frog's abdomen with a pair of bright, sharply pointed scissors and began working her way through the animal's innards, organ by organ, inspecting interstitial tissues and removing a number of items to test tubes for later histological examination. Liver, kidney, and heart tissue samples were deposited in either of two liquid-filled vials. One, headed for virology, contained a pinkish antibiotic fluid that would kill any bacteria present while leaving viruses unperturbed. The other, destined for bacteriological workup, held an amber-colored nutrient broth to promote bacterial growth. These assays and subsequent tests that would be performed on different animals—including parasitological examination and histological "sectioning," in which parts of the frog would be cut into paper-thin slices for microscopic inspection—would ultimately constitute nearly three-fourths of the lab's efforts to understand what was going on with the animals. At least two animals from each field site would be frozen as backup specimens, Meteyer said, "in case we get something significant in any of the frogs and need to do further bacteriological or virological isolation."

But this gross preliminary visual inspection was critical. Meteyer methodically performed the same examination on five frogs in a row, all with various degrees of ectromelia. On one, as Meteyer looked over the remnant of a partially missing leg, she detected the same kind of dorsal skin pattern disruption that Kathy Converse had commented on in the field when the animals were collected. Apart from the limb and pigment anomalies, the frogs appeared to be quite normal.

––––––––––

Bacteriology and virology were all well and good, but the attempt to link the deformities with a pathogen struck me as futile—little more than a kind of scientific due diligence. The possibility of an infectious agent causing abnormal limb development couldn't be excluded out of hand. Just about, though. Nobody could envision a mechanism by which a disease would target the formation and patterning of the legs. I was more curious about what the lab was finding with respect to a different kind of infection: parasites. The job belonged to a staff wildlife parasitologist named Rebecca Cole. I'd met Cole briefly in Duluth and had had a longer conversation with her at the Shenandoah conference. I was amazed at the passion she had for parasites, organisms with absolutely zero charisma. This fascination in turn bespoke a passion about life at its most elemental level. Pure biology. Unlike some other scientists I'd met, Cole wasn't much for subtlety. Life may be complex, but she seemed to think the science of life is about simplification, or should be. Cole also told fantastic, slightly creepy stories about doing fieldwork in the swamps of the Deep South— she'd gotten her doctorate at Auburn. At night in the swamp back then she'd push into a mossy bank and, pulling furiously with her hands, fill the bottom of a pirogue with snakes just to see what she could catch. Parasitologists, she said, found their subject everywhere. Roadkills were great, she added. "Always full of somethin'." More recently, at her home in Wisconsin, Cole had "lost" a large snake for a couple of weeks. Finally she'd spotted a few inches of tail sticking out of the waste drain in her laundry room and with some energetic tugging managed to extract the exhausted but still living animal.

"Don't ever tell that story," her husband, an airline pilot, had warned.

"Why not?" Cole asked innocently.

"Because if people knew that kind of thing happened around here nobody would ever come to our house again," he said.

I visited with Cole in the parasitology lab. She motioned to me to take a seat near a large table with a microscope on it. On the other side of the room I noticed a tall jar that was crammed with a great snarled ball of pearly-colored worms, each about the diameter of a finger and maybe nine or ten inches in length. "Came out of a wolf," Cole said offhandedly. "Pretty gross, huh?"

Cole said she was intrigued by the deformities problem, but skeptical about the depth of the investigation to date. Born and raised on a farm in Kentucky, she retained an Appalachian drawl that was disarming—Cole had a way of taking the edge off a question with an answer that invariably came back slow and resolute and unflinchingly direct. I asked her if I had it right that the Madison lab had received around seventy frogs from Minnesota the year before.

"Thereabouts," Cole answered.

"Did you examine all of them for parasites?" I said.

"No," she said. "A subset. About fifteen."

"And were any of those normal frogs?"

"Yes, but I think just one or two," she said.

I asked her what they found. Cole thought a bit before answering. She told me, for starters, that they had not cleared and stained any of the frogs as Stan Sessions had done with the frogs from Aptos, and that she regarded clearing and staining as a "nonstandard" technique for parasitology but one that seemed to be called for in this case because it had been used in the first instance by Sessions. For now, Cole said their examinations had been limited to careful dissections. The results were, in a word, inconclusive. This was slightly different from what Kathy Converse had told me a few months earlier, when she said they had found no correlation between parasitic cysts and deformities. Cole was telling me now that there didn't seem to be a correlation, but that they hadn't really nailed it down.

"In a general exam we did not see anything in the deformed frogs that was not in the control frogs," Cole said.

"Well, in terms of either the species of parasites or the way they were internally distributed, did you find any important differences between the deformed frogs and the normal frogs?" I asked.

"No," said Cole. "Again, the sample size was small. But there was nothing to me that was a red flag."

"So you didn't find anything really unexpected about the parasite loads in these frogs?"

"No."

"There was nothing in the way of a concentration of encysted parasites near the limbs that would suggest to you that they were associated with the deformities?"

"No," Cole said again.

Cole did say that they had recently dissected some deformed chorus frogs from Oregon that had extra legs and had in fact found some parasite cysts in the pelvic area of one. But the same region in another one of the frogs appeared to be free of parasites when the muscles were "teased apart" during dissection. Cole thought they might be able to "see deeper into the tissues" if the frogs were cleared and stained. But for now, she could only say that so far they had come across "nothing like what Dr. Sessions has seen."

I asked Cole about the idea that frogs were an intermediate host—a temporary stop—in the parasites' life cycle and that the amphibians' ultimate purpose in this sequence was to deliver the parasites to their primary host, a garter snake or some other carnivore that fed on frogs. She nodded emphatically. It was obvious that she appreciated the natural symmetries of such a narrative.

"Yes," she said. "Absolutely. Same thing happens with birds as the definitive host. It's a common strategy among trematodes."

But Cole was less certain about the rest of Sessions's hypothesis—specifically that through a coevolution over time parasites had been preferentially selected that would cripple the frogs in order to increase the probability of their being eaten by a predator. Cole said this sort of thing was known to occur in nature, that parasites sometimes really do alter the behavior or the physiology of an intermediate host to make it more susceptible to predation. A certain type of worm, for example, was known to infect isopods—small crustaceans—and change their color to make them more visible to the birds that are the definitive hosts for the worms. But Cole was doubtful that this was what was happening with frogs.

The problem was timing. Once the parasites entered the middle stage of their life cycle by leaving the snails and swimming free in the water they would have somewhere between twelve and forty-eight hours to find their way into a tadpole. This, Cole thought, was a relatively small window, although she conceded that different snails might "shed" the parasites at different times so that there could be an extended period when they were swimming about in a wetland. Even so, the time of initial infection would have to overlap the critical stage in the tadpole's development when the limbs would be sensitive to a disturbance. This was doubtless a similarly small window of time. Cole imagined that such precise timing in the wild would be a stretch, and she pointed out that parasites that did not infect the tadpoles at the critical stage would be passed along anyway.

"We do find these parasites in undeformed frogs," she said. "If they can cause the deformities at a certain stage, then that might play into making the animal more susceptible to predation. But I don't think it's necessary. I don't think it's necessary at all."

"Because predators eat a lot of frogs, including healthy ones that they are perfectly capable of catching without any advantage?" I asked.

"Exactly."

Well, then, I wondered, it didn't seem as if the parasites would have to cause the deformities "on purpose." That might happen purely by chance in those animals that got infected at just the right moment. Cole agreed, but said she was keeping an open mind. "I can understand how Dr. Sessions might reach the conclusion he did," she said. "I think it's a provocative idea and it raises some interesting questions. But it needs to be pursued to closure. What really needs to be done is to close the circle. You need to take the real parasites and the real tadpoles in development and let them come together and see what happens. That might be harder to do than it sounds, because the critical window might be only a couple of hours."

It seemed odd, when I thought back on this conversation, that I didn't see the problem here. But sometimes there's nothing like being firmly on the wrong track to keep you going. I had tried every question but the right question. Why hadn't they done more? Why did they stop after looking at only a few frogs? Whatever the reasons, it was clear in talking with Cole that the National Wildlife Health Center had not ruled out parasites as a potential cause of deformities in frogs.

The investigation went forward as if that was what had happened.

Many months later, I spoke with Cole again and this time I did ask why they hadn't done an exhaustive workup for parasites.

"Because the MPCA didn't ask me to," she said.

Cole said she thought that the cost of a larger-scale examination for parasites might have figured into it, and besides, the Minnesotans were also given parasite evaluations from Stephen Goldberg out in California. But she was adamant that the Minnesota researchers were warned that they could not consider Cole's results definitive. "I told them what we could do would not be statistically or biologically significant," she said. Yet when I pressed Cole on whether the MPCA had made a mistake in not pursuing the parasite angle more vigorously, she said she didn't think so at all.

"They did have limited funds," Cole said. "Why should they have thought that parasitology should get a priority over anything else they had to spend money on?"

I said I thought it would have been obvious: "How about 'Sessions and Ruth'? Wasn't there at least limited evidence that parasites could be a cause of limb deformities in frogs?"

"Aw, I don't think you can see it quite that way," Cole answered. "Look, this was a brand-new thing. They just wanted to see what was in some of these frogs in the way of parasites. And based on what I found I don't think you could argue that there was anything there that demanded more extensive work. This was just the first shot at taking a look for parasites. Actually, I think it was pretty amazing that they even did that much. Most of the time people don't think of pathogens or parasites when they see problems in wildlife. We see that overlooked all the time. It's just not the first thing that comes to mind."

It was evident, too, that the MPCA simply felt that parasites were an inadequate explanation for what was being seen in the field. The bulk of the deformities reported in Minnesota and elsewhere were not extra limbs. Sessions's growing insistence that parasites or predators accounted for missing limbs wasn't based on anything other than his own judgment that some natural cause had to be involved—a claim widely seen as an inherently weightless opinion.

———————

That same summer, as the Vermont countryside was being combed for deformities, the Minnesota inquiry lurched into its second full season of fieldwork with a long list of objectives and a new, fresh-faced young staff. Judy Helgen remained in control of the project, but she became an increasingly remote figure, routinely shunning the press and just as often playing hard-to-get with her collaborators—many of whom complained that they never heard from Helgen unless it was some sort of emergency. I talked to her from time to time, but almost always by prearrangement and never in the field. We met instead every few months in a conference room at the MPCA in St. Paul. These meetings were helpful, in a general way, but I rarely came away from an interview with Helgen with more than a smidgen of concrete information. Helgen was polite and sometimes even good-humored, but she was essentially impenetrable. Helgen deflected most direct questions about findings and work in progress and rarely offered insights of her own. She never brought actual data to an interview and typically responded to my requests for any kind of numbers or facts by saying something to the effect that she couldn't remember offhand but the information was probably written down somewhere. For the most part, I had limited success in getting my hands on any of it. Despite what seemed to be an almost complete absence of content in our infrequent conversations, Helgen insisted on taping them "for her records."

The day-to-day field investigation, meanwhile, was now led by Dorothy Bowers, a twenty-eight-year-old graduate student from the University of Minnesota who had until recently been working on frog population surveys in North Dakota. Bowers, quiet, short, and dimple-faced, wore bookish, oversized horn-rimmed glasses perched at the tip of her nose. With her pulled-back hair she looked like a casting director's idea of a librarian. In fact, Bowers was only a slightly bashful pit bull—bright and tenacious, an insistent workaholic who thrived on long days of grubbing about in the field and was equally driven in the office, where she endured a seemingly endless cycle of exhausting nights processing field samples and making ready to go out for more the next day. Bowers brought a focus to the Minnesota investigation that had been sorely missed. Where Helgen was constantly thinking up new things to do in the field almost literally from one minute to the next—her cell phone calls to field crews ordering changes in the daily agenda were legendary—Bowers made careful logistical plans and worked out exacting protocols for collecting frogs and environmental samples and shipping them off to the diagnostic labs. Back in St. Paul, Bowers undertook the daunting job of entering all of the existing field and citizen reports from previous seasons—none of which had yet been collated to date—into a computer database.

As the team leader in the field, Bowers was the chief agent in the MPCA's partnership with Jim Burkhart's project at the National Institute of Environmental Health Science (NIEHS) down in North Carolina. This meant collecting frogs at about fourteen sites—the number varied because of difficulties in locating permanently unaffected control wetlands—on a regular basis. It also meant collecting water, a lot of it, and sediments from which Burkhart hoped to determine some profound difference in the chemical makeup of the sites with deformities from the controls. This water chemistry analysis, much of which was done by contract at the College of William and Mary in Virginia, included an initial screen for several dozen "target analytes," various metals and chemical contaminants, especially pesticides, that could be quickly identified with standard assays. The metals list ran from aluminum to zinc, while the pesticides included a whole slew of common agricultural herbicides and insecticides, as well as some less widely distributed agents like methoprene.

In addition to sampling these "intensive study sites" repeatedly throughout the summer, Bowers was also helping coordinate a group of interns and temporary help from other departments at the MPCA that was conducting a "global survey" of deformities throughout the state. This grand-sounding project was arguably less comprehensive than it should have been. The survey team basically followed up on deformities reports that had been called into the agency, and along the way made a few random collections on public lands. This left out a large and potentially important part of rural Minnesota: farms.

I thought this was strange. Martin Ouellet's surveys up in Canada suggested a strong connection between farm pesticides and deformities. But Helgen told me that the MPCA had its hands full without looking for "trouble" in going onto private land. It wasn't clear what authority might be needed to do it, either, but that wasn't the direction Helgen felt the investigation should go anyway. The MPCA, she said, "can't really change what it's doing." Unless it would be to narrow the investigation even more. This, Helgen said, was actually a good idea: "We're at a point where we feel we have to focus in on a few sites. We can't assess the scope of [the deformities outbreak] the way you would ideally like to do by randomly picking sites in different land-use areas. We just can't take that on."

That was easy enough to believe. What the MPCA was already doing had extended everyone to their limits—the ceaseless frog collecting and data-taking, the water sampling and shipping, the never-ending days of driving from one site to the next, thousands and thousands of miles logged crisscrossing the state over and over again, coming back time after time to St. Paul long past dark, wrecked by wind and sun and swampwater, but with hours of work left to do—it was too much. By midsummer Dorothy Bowers had accrued enough comp time to take off the rest of the season had she been of a mind to do it. The Minnesota deformed frog investigation, which only a year earlier had been an anxious, peppy blitz, was becoming a forced march across the marshes. A year earlier everyone speculated on how closely an answer loomed. Now Bowers thought there might be only the oblivion of permanent uncertainty. By the end of the summer, she wasn't sure that what she had seen meant anything at all.

Bowers was troubled by the lack of clear evidence delineating either the scope or the severity of the deformities. Only a handful of sites exhibited truly distressing rates of deformities, and even these numbers sometimes gyrated from one visit to the next. Most of the data fell into a gray area. Some of the affected sites had deformity rates of less than 2 percent, while some of the supposed control sites had rates approaching 1 percent. What was the difference? Deformities appeared overnight in places they weren't before and disappeared from sites where they'd been previously detected. What did this mean? Bowers fretted that they might be cataloging normal variations and not much more. It's nature. Stuff happens. Even in the places where something seemed clearly amiss the signals were mixed. Several sites had deformity rates of around 5 or 6 percent—outside the presumptive baseline normal range but not nearly so dramatic as the sites where the rate pushed above 20 percent. And nothing at all looked like the Bock's lake, CWB, where well over half the animals in one species were deformed. Bowers thought it was impossible to see a pattern in any of this, and she told me at the end of the season that fall that she was unsure there was really anything to investigate.

We met at an Indian restaurant north of St. Paul and drank a couple of beers in the wan twilight of an afternoon in November. The fading light was sullen and Bowers was hesitant—she didn't want to say anything to undermine the project she'd worked so hard on. But I could see she had major doubts.

"I think the data has been kind of misconstrued a little bit," Bowers said finally. "I mean, we keep quoting these huge numbers of reports of deformities from across the state. But I think that can be easily misinterpreted, because we go out to these sites, and you know what? We don't find very high percentages of deformities. It *sounds* like there's a big problem. But when you go out and you look at the places where you've got more than ten percent abnormalities there are only . . . what? fourteen places in the whole state? That makes it a totally different scale of a problem to me. We haven't gotten anything this year that gives us an idea of what kind of problem we're really dealing with. And we don't have any historic information because we just haven't done this kind of survey before. I think the whole deformed frog problem is kind of premature. Maybe it's not as much of a problem as everybody thinks it is. Maybe it's been blown out of proportion."

She took a breath.

I said I didn't get it. I asked her if she might not be seeing things in reverse. To me, it was the contrast that was telling. Wasn't it the overwhelming normalcy that prevailed almost everywhere that demonstrated the seriousness of the problem in those places where the deformities occurred? In more than a hundred visits to almost eighty different sites, the global survey team had turned up only five new hot spots with significantly elevated rates of deformities. I told Bowers I had been to one of these, near a large lake in Chisago County in east-central Minnesota, where about 12 percent of the leopard frogs were deformed. I'd felt that was pretty convincingly outside the norm at the time, but felt even more strongly about it a few weeks later after spending two days on the road with the global survey team and checking out one place after another without finding anything. Hot spots were hot precisely because they were departures from the norm. Plus, I added, if you've got fourteen legitimate hot spots there are for sure plenty more out there, because there's no way you can see everything. *Nobody* can get to every single wetland in Minnesota.

Bowers was unmoved.

"But why *don't* you expect to find deformities from time to time?" she said. "I mean, why isn't that normal?" She looked out at the purpling sky. It was darker inside the restaurant now and her face was reflected in the glass of the window. Bowers's mood was quite different from what I'd seen when she was in the field and all business. She seemed worn out by it all and almost compulsively mindful of the many holes in the deformities story. If you wanted hard proof of a widespread outbreak of deformities in Minnesota, she said, the efforts to date had not accomplished it. "To me," she said, continuing to stare at the sky, "nothing has been excluded. I'm saying it could be something biotic. I'm saying it could be something else. I don't think it's been proven that this is not a natural occurrence. And I don't think it's been proven that it's not caused by a chemical. It may be a chemical."

"But what do you think is happening?" I said. "Anything?"

"I don't know," she said.

———————

That was, of course, exactly the concern. Nobody knew what was going on with Minnesota's frogs and everyone realized that all possible causes for the deformities had to be given serious consideration. One agency that was watching these developments and trying to gauge the uncertainties from the sidelines was the Minnesota Department of Health. When Judy Helgen had told me in Duluth that the day might not be far off when some sort of public warning might have to be issued about the safety of Minnesota water, the mere suggestion had sounded almost unthinkable. But the possibility was real enough that someone was thinking about it.

His name was Hillary Carpenter. Carpenter was an environmental toxicologist in the health department. When I talked with him that summer, he told me he handled a variety of "general toxicological issues," but that more recently he'd become the agency's designated endocrine disruption specialist. Carpenter said he'd first heard about the deformities like everyone else when the initial outbreaks were reported in the local papers. Shortly after that he'd attended an in-house briefing by Judy Helgen over at the MPCA. Carpenter recalled that there was talk about an imminent "deluge" of phone calls, from citizens who were going to be worried about possible human health consequences associated with the deformities. That really hadn't materialized yet, he said, but the health department had remained in close contact with the investigation ever since and had in fact been giving some thought as to how a human health threat might be handled if one came to light. Carpenter said he'd been skeptical at first that this would turn out to be anything more than an isolated wildlife problem. But more recently he'd become convinced that the possibility of a chemical involvement was very real and very worrisome.

"When I first heard Judy talk about what had been found at the Ney farm I wasn't too sure this was something that would even turn up anyplace else," Carpenter said. "Then as things spread out more we got more interested." Carpenter said he'd gone to the Duluth meeting "waffling" on whether there was any kind of human health concern at stake and had come back even more doubtful of it.

"So Stan Sessions convinced you it was all natural?" I said.

"Well, yeah," Carpenter said. In spite of Sessions's last-minute reluctance to attribute all of the deformities in Minnesota to parasitism, Carpenter said he'd come away believing that parasites sounded like a better explanation than an unknown chemical that was somehow randomly contaminating isolated wetlands without rhyme or reason. Carpenter admitted that his reservations were owed at least partly to the fact that a chemical cause was simply too alarming to believe. "It appeared that you'd have to invoke some theory of atmospheric transport to get a chemical to all these places where we wouldn't otherwise expect to be seeing it," he added. "And that was a frightening thought, because we don't know much about materials like that being transported through the air."

But Carpenter had a change of heart after the Shenandoah meeting the following spring. He said he became "much more convinced" that the cause of the deformities was chemical and a lot more dubious that parasitism or predation were extensively involved. All of the arguments that could be made against chemical causes became "less acceptable," he said, after listening once again to the explanations offered by Ken Muneoka and Dave Gardiner at Shenandoah on the role of retinoic acid in limb development and the extensive experimental evidence that too much or too little of the compound could produce deformities very similar to what had been seen in the wild. Carpenter said he'd been persuaded that "it's gotta be retinoic acid" or something that mimicked the powerful hormone. This meshed perfectly with the idea that the deformities resulted from endocrine disruption, just as Pat Schoff had first proposed to Joe Tietge almost a year earlier.

Carpenter said it was painfully obvious that Gardiner and Muneoka had gone to Shenandoah hoping to sink Stan Sessions and the parasite hypothesis. Carpenter felt they had succeeded. "I sensed a definite rivalry there," he said. "I think they felt that Sessions was maybe winning everyone over and they had to bring things back into perspective."

Perspective on the deformities, however, still seemed to shift continually. As impressed as Carpenter had been by the theoretcial underpinnings for a chemical cause of the deformities, he said he still felt the outbreak in Minnesota was poorly understood and that in speculating on its causes people were getting ahead of themselves. He thought Martin Ouellet had made a strong case at Shenandoah that the Minnesota investigators hadn't really done a comprehensive survey of the state and were instead describing the whole deformities problem only in terms of hot spots. Carpenter believed that Ouellet had been disappointed with the U.S. data at the Duluth meeting but had "swallowed his tongue" back then in the belief that the Americans just needed more time to get on the right course. "Martin thinks that the way we're looking for deformities only ensures that we will find them," Carpenter said. Monitoring hot spots was a kind of self-fulfilling proposition. Ouellet insisted that what was needed instead was a broad, random survey that would put together a statistically valid profile of wetlands in a way that was more representative of what was in the wild. Carpenter agreed. There was no clear picture of what was going on in Minnesota, he said, and the dearth of "hard evidence" made any kind of definitive statement about a deformities outbreak impossible. The MPCA's decision to focus on just over a dozen sites in their intensive study, Carpenter added, was a "biased approach that just means you're going to see deformed frogs."

Those reservations notwithstanding, Carpenter indicated that the health department was not sitting idly by until more persuasive data came in. To my surprise he said the agency had already considered how it might issue public warnings about potential human health hazards in the event they were necessary. Carpenter said they would most likely do something similar to what they already did with regard to contaminated fish: They would issue advisories against exposure to polluted waters. Such advisories are well known to Minnesota sportsmen and are widely posted at boat landings all over the state.

"Minnesota has always been conservative on these matters," Carpenter said. "We have always erred on the side of protecting people. For twenty years now we've routinely issued warnings about consuming fish in areas contaminated with mercury or PCPs, with special emphasis on sensitive populations like pregnant women and children. We recommend that those groups not eat fish more than once a week. That's because we feel strongly there's a potential problem there. I think the same kind of conservative thinking would apply if it becomes probable that these deformities are the result of endocrine disruption by a chemical contaminant. This might be seen as real evidence of effects that would carry over to human health."

What those effects might be was anybody's guess, Carpenter said. But he agreed with Pat Schoff's assessment that deformed limbs in human babies could be a remote possibility because of the likelihood that a fetus with such abnormalities would not advance to full term. This was far from certain—after all, fetuses exposed to thalidomide had been born with horrific limb defects. Carpenter indicated he would tend to worry more about an effect on humans that was masked by early terminations of pregnancies.

"I'd guess the chances for a multilimbed fetus to survive would not be good," Carpenter said. But the odds of ever knowing that one had been conceived were equally low. "What we might expect is a decreased rate of conception or an increased early pregnancy failure rate. Or both. And the problem is that you're already looking at a fairly high baseline rate for that sort of thing. We didn't know this until recently, but the failure rate of pregnancies is about one in three." This isn't routinely apparent, Carpenter said, because the vast majority of spontaneous abortions occur so early that the woman never realizes she is briefly pregnant. If something that was causing deformed frogs also elevated the rate of miscarriage in humans even marginally it would be disturbing—but possibly so subtle as to avoid detection.

On the other hand, Carpenter said it was important to consider the possibility that humans simply wouldn't be exposed to a contaminant the same way a frog is, and there might well be no effect at all on people.

"When you're dealing with wildlife," he said, "you're looking at populations of organisms that tend not to be very mobile, that tend to select an environment and stay there, and that have certain foods that they eat all the time. So their exposure to something in their surroundings or their food is more or less constant. Humans aren't like that. We move around a lot. We eat different things. We have a very different type of reproduction. For years it was thought that external chemicals could not have any effect at all on human fetuses because of the placental barrier. Now that turns out not to be absolutely true. But certainly it does afford some protection."

Carpenter thought that if his agency did have to act in the deformities problem that they could anticipate strong opposition. Politically and economically, the consequences of a broad warning about the safety of Minnesota water or croplands could be devastating. But the department had weathered such opposition in the past and could do so again.

"Remember," he said, "in Minnesota it's considered almost a birthright to fish and to eat fish. When the Department of Health had to say to people, 'Don't eat fish', that had a big impact. It affected a lot of citizens and it negatively impacted tourism and fishing guides and the charter industry. There were some real severe political repercussions. Judy Helgen has already made the point that there are very few reports of deformities coming in from the far northwestern part of the state and that the politics of the agricultural economy in that area were such that it's unlikely we'd get reports even if they had deformities up there.

"Traditionally, the chemical industry is not real willing to admit to causing health problems. But through it all the state has stood behind the health department and said we were doing the right thing. And I suspect that would happen if we determined that there was an endocrine disruptor involved in the frog deformites. I think we'd be very aggressive. If we determine that we're dealing with something that causes birth defects or reproductive problems we're going to act immediately."

I asked Carpenter if that time were close at hand.

"I don't think so yet," he answered. "I don't because the evidence is so shaky. It's something to be concerned about and needs to be followed closely. My message is always that yes, this is a cause for concern. Otherwise we wouldn't be involved in it. But do we have enough information at this time to make any definite statements about effects on human health? No."

When I asked Carpenter how the investigation was going he answered that it seemed to have taken on a life of its own. He didn't see any end in sight. He said he called Judy Helgen often for updated information, but that he rarely heard back from her.

Dorothy Bowers quit a few weeks after we'd talked, in November of 1997. She moved to Washington State. I thought all winter about what she had said about how the deformities problem seemed to blur around the edges the more time you spent looking into it. The uncertainties could be disorienting. I recalled one stop in particular on my tour with the global survey team that drove the point home. It came in the middle of a grueling day, when we left St. Paul just after dawn and ended up in a dingy motel two hundred miles to the north just before midnight.

The site was a residential location, a house in the country a few miles from the town of Kensington in west-central Minnesota. Kensington is locally famous for a stone discovered there in 1898 upon which were carved supposedly ancient Norse runes that were interpreted by some as proof that Vikings had visited the area on a trip to North America more than a century before Columbus. The Kensington Runestone has long since been dismissed as a hoax, but a whiff of strangeness lingers in the town, and I thought I could feel a little of it when we pulled into the large yard in front of a small, neatly painted blue ranch house. There were children's toys scattered about near the house, and one of the older of the four kids who came outside to greet us with their mother said he'd caught a number of frogs with missing or partial legs that summer. Also some abnormal salamanders. Whereupon he jumped down into one of the basement window wells and dug around with his hands for a minute before producing two lively looking salamanders. Both of them were normal.

We spread out over the lawn, which sloped for a hundred yards toward an old barn with beautifully weathered timbers, and spidering gaps in the siding that were shot through with sunlight. Marshy-looking areas were all around the property, and there were more frogs than you could believe immediately underfoot. Juvenile leopard frogs exploded across the grass in ground-level tsunamis as we moved away from the drive. I actually worried about stepping on them, there were so many. Four of us stutter-stepped around for about fifteen minutes, sweeping our nets back and forth in front of the fleeing animals. When we stopped we had caught 160 frogs.

It took an hour or so to look them over. Only three had any kind of limb irregularity, and these were fairly minor problems of the foot and lower leg. Three out of 160 is just under 2 percent. Depending on how you looked at it, there was nothing going on here or this was a site having almost double the normal baseline rate of deformities.

Which way would you have it? This was one of those perfect, biological moments, when the whole volatile flux and funk of life itself was vividly self-evident. On more than one occasion Martin Ouellet had told me that anyone who ventures into the field with a question in mind will soon enough find a different problem staring them back in the face. "If you look hard enough," Ouellet always said, "you will find *everything*."

Later that winter I had one of my scheduled chats with Judy Helgen and I asked about whether she'd felt a shift in what they were seeing. She agreed that at the tail end of the 1995 season and then, in 1996, it seemed that there had been more places with really astronomical rates of deformities. The possibility that some of these hot spots might have cooled slightly didn't sway her much. "My overall impression is that it's not much different now, other than we're not seeing some of the extreme rates," she said.

"So what do you make of what is still being found out there?" I said.

"Well, I wish we had an answer," she said.

"Do you think this is a real problem?" I said. "I mean still?"

"Yeah. Yeah," she said. "I think it's a real thing. And it's obviously not just going to drop away."

It followed that if the problem persisted, then sooner or later there would be an explanation in hand—probably more than one. I suddenly wondered, as I listened to Helgen going through some of the water chemistry studies that were under way, what would happen when that day came.

"Judy," I said, "let's say that someday you find something in the water from the affected wetlands that has some sort of developmental toxicity to it. Then what do you do?"

Helgen smiled. "Tell everybody," she said.

Bad Weather

ONE GENERAL TREND SEEMED TO HOLD UP: MOST OF THE DEFORM-ities reports came from the northern part of the country. This fueled speculation among the researchers that there was a geographic factor involved in the deformities. Among the regional influences that might be considered were the deposition of airborne contaminants by prevailing winds, a water "matrix" in the North that was in some way different from the one down south, larger changes in atmospheric penetration by ultraviolet radiation at higher latitudes, and differences in land-use patterns and pesticide applications. None of these possibilities had thus far advanced beyond the conjectural stage, but the observation that deformities were less common in the South remained a puzzling part of the backdrop of the investigation.

It also presented a sticky question for proponents of the parasite hypothesis—which Stan Sessions had now explicitly married to the idea that predation contributed to the missing-leg phenomenon. Reckoning that he could no longer ignore the fact that frogs with missing or partial legs far outnumbered multilegged frogs in the deformities reports, Sessions had grown more insistent about his dual contentions that parasites might block limb development in certain instances, but that predation was the

primary cause of missing legs. Sessions became adamant that limb amputations were a routine feature among amphibian populations and that such
"deformities" were not developmental in nature. Missing legs on frogs,
Sessions said, were instead only an example of a common trauma that was
often observed by "experienced herpetologists."

But if that were true, then why didn't missing limbs turn up wherever frogs and predators lived together—including the southern United
States? Nobody spent much time trying to refute the parasite/predator
hypothesis, mostly because nobody believed it had serious merit. On its
face it was a two-headed proposition, neither part of which enjoyed much
support, especially the predation part. Predators weren't confined to hot
spots—they were everywhere—and to suggest that they were somehow
more active in the very same wetlands where parasites were causing other
kinds of deformities struck most people as ludicrous. If predation caused
missing legs, then missing legs should have been turning up everywhere
and for as long as people had looked at frogs. That would include the
southern United States. In fact, the phenomenon ought to be relatively
common throughout the world. Sessions's claim that any good herpetologist should know this rankled a lot of good herpetologists, among them
Dave Hoppe, who grew agitated any time Sessions's name was even mentioned. Hoppe invariably referred to Sessions as "Stanley," and usually
only to point out that, in his view, Sessions didn't know what the hell he
was talking about.

What were these supposed predators, anyway? Mike Lannoo, who'd spent
two decades at Okoboji looking at every creature that lived in the wetlands,
told me he could not think of anything that preyed on amphibians that was
capable of amputating legs except in rare accidental "near misses," which
would never turn up in large numbers. Like Bob McKinnell, Lannoo maintained that predators were almost always after the whole frog, not just part
of it. Anything that got hold of any part of a frog with the intention of eating the animal usually got what it wanted. Larger predators like birds or
small mammals that attacked from above, said Lannoo, would find a developing leg on a tadpole a very small, very out-of-the-way target. Predators
that might attack from below—from underwater—seemed an even further
stretch to Lannoo. Small fish and some of the bigger insect larvae might
conceivably nibble at tadpole legs as they trailed in the water next to their
tails. But none of these animals, in Lannoo's estimation, had the requisite

"shearing action" in their bites that would produce the neat, scarless removal of a leg that would be consistent with what was being found after the fact. Bigger underwater predators, Lannoo added, were not going to be satisified with a dainty morsel like a leg. At a minimum, you'd expect to see tadpoles with missing tails more often than missing legs. Besides, lots of midwestern frogs—including leopard frogs—-much preferred to breed in small, shallow ponds that didn't harbor larger aquatic predators, especially fish.

I had a chance to see what Lannoo meant one day in the summer of 1997 when I visited him at Lakeside Lab in Okoboji. Lannoo and his evolution class had repaired to a wetland one afternoon and were pulling a seine through it when they caught several larval *Ambystoma tigrinum*—tiger salamanders. Lannoo was particularly interested in a couple of specimens whose large size, big mouths, and blocklike heads indicated they were a variant called the cannibal morph, so named for its ability—and willingness—to feed on other tiger salamander larvae. Not to mention anything else that swam within reach. Lannoo was something of a cannibal morph specialist. Although the cannibal morph was known out west, Lannoo had been the first to discover it in Iowa.

Lannoo brought several animals back to the lab that afternoon, including a good-sized cannibal morph salamander. Its thick, lizardlike body was roughly ten inches long. Lannoo brought out a small aquarium, filled it with water, and set it on a large, square wooden table in the middle of the classroom. He added the salamander, which wiggled in slow circles briefly before coming to rest on the bottom of the tank. He asked the students to take their seats for a discussion of what they'd seen in the field that day. While they talked, he said, they would watch "a small experiment."

As everyone sat down, Lannoo brought out a leopard frog tadpole from one of the big cement specimen tanks lining one side of the room and dropped it into the aquarium. The tadpole looked to be around three inches from its snout to the tip of its tail, about a third the size of the salamander—which acknowledged the presence of something new in the tank by beginning a steady upward drift without so much as a twitch of muscle. The tadpole swam about, oblivious to the dark shape rising stiffly toward it from below.

Lannoo began a review of the plants and animals they'd identified that afternoon. The students dutifully recounted their taxonomic mantras of family, genus, and species. But no one could take their eyes off the aquarium. The tadpole continued to swim aimlessly about, but the salamander steadily closed the gap until the tadpole was literally brushing alongside it. This went on for a few minutes. Neither animal showed the slightest obvious interest in the other.

Then there was a small explosion in the water, over before you could see just how it happened, and the tadpole was left madly thrashing for its life. Only now it wasn't going anywhere. Its right hind leg was firmly clamped in the salamander's powerful jaws.

"Now here's an interesting scenario," Lannoo said. "Let's watch for a minute."

The tadpole jerked and twisted violently. The salamander held on fast, its thick head barely moving as the smaller animal struggled mightily to get away. It seemed the leg would surely part company with its owner. But it didn't. The salamander held on. Then came another blur of amphibian flesh. In a single, rapid movement the salamander shifted its bite. Now the tadpole was held crosswise, with the salamander's jaws clenched around its torso. The tadpole's stuggle was suddenly much reduced. The salamander drifted slowly forward with its prey. After a while there was another violent readjustment and the tadpole was then headfirst in the salamander's mouth. The tadpole no longer moved at all yet seemed somehow to be swimming down the salamander's throat. Little by little it disappeared. In a minute it was gone. Just like that. A routine example of predation, the kind seen by experienced herpetologists all the time. The salamander sank back toward the bottom of the aquarium.

In late June of 1997 I drove over to the St. Paul campus of the University of Minnesota to meet Dave Hoppe. Hoppe was beginning an unusual frog survey he hoped would shed more light on what the normal "background rate" of limb deformities was in frogs in Minnesota. His own experience—that such abnormalities were quite rare—wasn't enough. He wanted to double-check the data, and because of the way things used to be, he could.

In the old days—before scientists became worried about population declines among amphibians and before animal-rights proponents had begun to shadow every move made by academic researchers—it was common practice among field herpetologists to collect and preserve large numbers of animals for no particular reason. Many such collections are cached in university biology departments all over the world. There's a pretty good one at the University of Minnesota.

The collection is stored in the basement of the Ecology Building. The frogs—there are thousands of them—are kept in big jars marked with the species, the place where they were collected, and the date they were caught. All of them had been collected by David Merrell, a now-retired university biology professor who for many years had been the state's foremost herpetologist. Hoppe had actually taken classes from Merrell as an undergraduate and had recently stopped in to see his old professor to ask him about the collection. Hoppe was particularly interested in how Merrell had chosen which frogs to keep, especially whether he had routinely tossed away abnormal ones for being of no interest—a possibility some researchers had raised of late.

Merrell told Hoppe he didn't choose at all. He had kept everything he caught when he was collecting and returned all the animals when he wasn't. Even when he wasn't keeping frogs for his collection, Merrell said, he paid close attention to anything that looked out of the ordinary. In fact, Merrell added, he had indeed observed a large number of deformities on one occasion and had even taken the time to publish a report of the discovery. Hoppe was amazed. Did Merrell by any chance have a copy of the paper?

He did.

The title of the report was "Natural Selection in a Leopard Frog Population." It had appeared in 1969 in the *Journal of the Minnesota Academy of Science,* and because of the relative obscurity of that publication—plus the absence of any mention of deformities in the title—the paper had gone undiscovered by everyone working on the current frog problem. But Merrell's observations from almost three decades ago bore an uncanny resemblance to what people were now finding in Minnesota.

In the summer of 1965 Merrell had been engaged in population studies on leopard frogs at a number of locations in the vicinity of the Twin Cities. On a visit to one of these sites just north of Stillwater, Minnesota, in July 1965, Merrell was astonished to discover eighteen deformed juveniles among the eighty frogs he caught that day—a 22 percent rate of abnormalities. The deformities included abnormal toes, missing portions of the lower legs, and, in a few instances, legs that were completely absent. All of the deformities were in the hind limbs, and all were "unilateral," affecting only one side or the other. Merrell observed that the deformed animals moved surprisingly well and seemed to be normally active, though he felt sure they would make easier targets for the garter snakes that were feeding hungrily on young frogs at the site.

If Merrell had more than a passing curiosity about the causes of the deformities, he didn't indicate as much in the paper. Instead, he focused on the fact that the deformities were found in juvenile frogs and that their frequency decreased as the season progressed—which suggested that missing legs were maladaptive and this phenotype would therefore not survive in the population. But embedded in the narrative of the report were a number of extremely telling observations that seemed to bear directly on much of the current discussion about the deformities now being investigated.

Merrell noted, for example, that some of the frogs he caught showed evidence of having been "wounded." These injuries, he wrote, could be attributed to predators, cars, or much more rarely to the animals' capture in the net. But Merrell didn't think it was hard to distinguish trauma from a developmental problem. "The differences between wounded and deformed frogs were so clear-cut that no hesitancy was felt in scoring one or the other condition." Merrell recorded wounded animals as "normal." Natural selection, the evolutionary process referred to in the title of the report, usually pertains to differences in surival rates among animals with slightly different genetic profiles. When one phenotype is "selected" because of a small advantage it has gained in the wild its genes are passed on to subsequent generations—and the phenotype thereby becomes more prevalent over time. Merrell in this case used the term *selection* loosely, meaning only that deformed animals would not survive to reproduce. Merrell was actually convinced the deformities were not the result of a genetic mistake because by the time of metamorphosis all of the progeny of many different matings in the wetland would have been "thoroughly and randomly mixed" so that the sample would have included progeny from many different egg masses. This, coupled with the asymmetry of the deformities, Merrell reasoned, suggested the cause probably had a "large environmental component."

Merrell didn't speculate what that factor might be, but he did hint that he thought the whole episode was a pretty rare event. Despite the "wealth of genetic variability" that occurred in frog populations across a large geographical region, Merrell observed, a leopard frog almost always looks like any other leopard frog. The standard phenotype, he wrote, was "uncommonly common." In his final collection at the site that year, on a blustery day in late September, Merrell found the rate of deformities in his sample had dropped to just over 3 percent. However, this was, he noted, "still far greater than that found in other populations." When Hoppe asked Merrell if he'd ever seen anything like this anywhere else in all the years he collected frogs in Minnesota, Merrell said he had not. Hoppe handed me a copy of the Merrell paper as soon as I arrived. "Check this out," he said.

The jars full of frogs in the basement of the ecology building had been collected in 1958 and 1959. When I got there Hoppe was inspecting a small pile of leopard frogs. Next to him on the lab table were several large jars packed with more of the same. Hoppe was being helped by Erik and Larissa Mottl, two of his students who'd graduated that spring. The husband and wife team had worked briefly for the MPCA but left after a dispute over hours and pay. They were happy to have landed work for the summer with Hoppe.

A faint odor of alcohol filled the room, though it was much less noticeable than you might expect given the open jars of forty-year-old pickled frogs sitting around. The animals looked suprisingly well preserved, even a little lifelike. Their color was only slightly faded, their bodies stiff but not completely inflexible—about the same consistency as a rubber Gumby. Erik removed the animals one at a time, measuring them and noting any peculiarities, then passing them on to Hoppe for closer inspection. Larissa recorded snout-to-vent lengths and any abnormalities that were found, though it was already apparent that there were not going to be many. All the frogs, representing a mix of ages and sexes, looked pretty normal. Hoppe said that there were more than enough frogs in the collection to establish a reasonable baseline rate of deformities that could then be compared to a new re-survey of the same sites. This was something he intended to begin this summer, he said.

"Whoa!" Erik said as he plucked another frog from a jar—an adult with two extra hind legs. They were sticking out of its mouth. "Here you go, Dave," Erik said. "This is just what you said we'd see sooner or later."

Hoppe laughed. This was actually very interesting data, he explained. He had often noticed that adult leopard frogs, which normally spend the summer widely dispersed across the landscape, tend to recongregate near their breeding ponds just as the young of the year are metamorphosing and emerging from the water. "I've always felt fairly certain they were coming back to eat the kids," Hoppe said.

Sure enough, as Erik gently pried open the big frog's mouth with the forceps, the half-swallowed smaller frog attached to the legs sticking out was plain to see. "Hello in there, Junior," Erik said.

"Gross," said Larissa.

I asked Hoppe about his site in Crow Wing County—"site" being the only way I could describe it. In his presentation at Duluth, Hoppe never gave specifics about CWB or the Bocks. It was all a closely guarded secret, known only to Hoppe, the Mottls, and the MPCA—which wasn't saying anything about any of its study sites anymore. Judy Helgen, afraid the landowners she was working with would back out of their agreements to let the agency conduct studies on their properties, had ordered a total blackout on the dissemination of information—especially the locations of deformities hot spots and whatever was being found at them. Earlier that spring I'd met her and Mark Gernes down at the Ney farm—a place I'd visited on numerous occasions—and she'd made it clear that she felt the site had been contaminated by too many visitors already. She was determined not to let that happen at other study sites. She said that it would be a great misfortune if anyone were to learn of the location of a site like the one where Dave Hoppe was working. So I wasn't sure what Hoppe would or would not say about his work up near Brainerd.

"Carry this for me," Hoppe said, handing me a jar of frogs that had been examined and were ready to go back on the shelf. He picked up another and we walked through a doorway and into the cavernous room adjacent to the lab where the specimens were kept. The Merrill collection occupied several large gray steel lockers. Hoppe opened the doors to one of them and carefully replaced the jars. Behind us were row upon row of tall open shelves, each lined with more glass jars as well as a fair number of big plastic buckets—all well sealed—each of which contained specimens of various amphibians and reptiles that had been collected throughout the Western Hemisphere by different university herpetologists. Some of the jars and buckets were quite large, as, presumably, were their contents. Hoppe was silent for a moment as he shut the cabinet doors, leaving the frogs behind.

He'd just checked in with the landowner up in Crow Wing County, he said finally. The kids there had been out a couple of days ago and managed to catch ten mink frogs. Nine of them were deformed. Hoppe said the kids had found extra legs and also skin webbings on the hind limbs. He said he thought the peak of metamorphosis was almost at hand for mink frogs and he was going to be heading up there in a couple of days to see what he could collect.

I asked Hoppe if I could meet him there and started explaining that I only wanted to observe his fieldwork and I wouldn't get in the way or disclose the location or anything about it without his permission. Hoppe waved me off.

"Do you own waders?" he asked.

I told him I had both hip boots and chest waders.

"What are you, a duck hunter?" he said.

"Sure," I said.

"Bring 'em both," he said. "We'll meet you at the Brainerd YMCA at eleven A.M."

The twenty-eighth of June was hot and overcast. Brainerd is about a two-and-a-half-hour drive north of where I live outside the Twin Cities. It's not much closer to Alexandria, the small resort town in the western part of the state where Dave Hoppe lives. I left early and got to Brainerd with enough time to spare to sit on the tailgate of my truck and read another David Merrell article, his monograph on the natural history of the leopard frog. A couple of observations stood out. Merrell wrote that leopard frogs were a favorite lab animal with parasitologists because of the "variety and abundance of parasites they harbor." But in the same passage, Merrell added that dead or diseased leopard frogs were almost never observed in the wild.

Merrell also wrote that leopard frog tadpoles will try to swim down into submerged vegetation to escape danger, but that this was really not much of a defense. Their only real protection during development, Merrell noted flatly, was "the choice of a suitable breeding pond by the parents." In normal circumstances, that meant a wetland too shallow for fish but deep enough that it wouldn't dry out before the tadpoles reached metamorphosis. But it was by now obvious that something else was making many Minnesota wetlands inhospitable places for frogs.

Hoppe and the Mottls wheeled into the parking lot behind the YMCA at exactly eleven. Hoppe, I soon learned, is always on time. We drove out to the Bocks', east of town, turning off the county road and onto the long driveway that led up to the sunny yellow house. Buster, the Bocks' friendly, overweight golden retriever, woofed his way over to us as we got out of our vehicles, wagging his tail furiously. Hoppe asked Erik to put the dog in his fenced-in kennel. "Buster likes to catch frogs," he explained. "I can't let him come down to the water when we're collecting or we'll have deformed frogs with puncture wounds in them, too."

Everyone pulled on hip boots. Hoppe assured me we wouldn't be working much more than knee-deep, though I'd want the chest waders for later in the day when we went to the nearby lake that was the control site for this location. My wading boots were warm from riding in the bed of the truck and the air around us was heavy. Rhonda Bock came out with Brandon and Troy, who were eager to tell Hoppe about the frogs they'd saved for him in the cages he'd left down at the beach. Dennis had worked the night shift the day before and was still sleeping. Rhonda said he'd probably come out later on. Hoppe had long-handled nets for us to use. Erik took two, saying he'd perfected a technique for shepherding frogs into one net with another. We each also took an old pillowcase from the back of Hoppe's Jeep—to sack frogs as we caught them—and set off across the yard and down the hill to the water.

The lake lapped at a strip of brown sand at the foot of the Bocks' yard. The empty dock stood out forlornly from shore, the paddleboat beached next to it. Beneath the patchy sky, the surface of the lake looked blank but inviting in the heat. The sun came and went from behind the clouds, and vagrant puffs of the wind pushing at the treetops swirled downward with just enough force to keep the bugs away. CWB didn't look special or poisoned. Maybe it wasn't. Maybe it was. Rhonda, who'd brought some lawn chairs down by the lake to watch the work, told Troy and Brandon they could help along the beach, but to stay dry. "I don't want you in the lake," she called out.

We waded in. Larissa moved off to the right, to hunt frogs along the shorter southern edge of the propery. Hoppe, Erik, and I went left, into the tall reeds and matted islands of vegetation that followed the bank for a hundred yards or so up to the old fenceline separating the Bocks' from the dairy farm next door. Hoppe swung out along the edge of the weeds; Erik and I slogged ahead through the shallows, churning up clouds of muck. The bottom was very soft; I could feel the mud pressing upward along my booted shins as I sank in repeatedly, one leg at a time. We had to move slowly, in the way wading outdoorsmen have done forever, being careful to get a foot loose for the next step before leaning forward to take it. The way you fall over in waders is by trying to walk with both feet trapped in the bottom. Nobody wanted to fall in.

. I didn't have any experience with mink frogs, and Erik patiently explained how to spot them. Look directly into the water, he advised. Erik shot an arm forward and brought up a few pounds of muddy vegetation that dripped darkly from the bottom of his net. He reached in and extracted a mink frog coated in grime and held it out for me to see. "Oh yeah," Erik said. The young frog still had a long tail, but its well-developed hind legs were heavily webbed behind the knees and a paler spike of a limb grew out from one side ahead of the primary leg. Erik pulled his pillowcase loose from a belt loop, dropped the animal in, and then pushed ahead again slowly. "Watch for their eyes and then try to spot the body hanging underneath," he said. Erik again thrust his arms forward, this time scissoring the two nets together. Up came more mud and another frog.

To catch frogs, Hoppe explained to me one time, you have to look only for frogs. He meant you had to be able to exclude everything else that crossed your field of vision. Hoppe said it was a matter of developing a "search image." Like a predator. You pay attention only to things that indicate frog—a pair of eyes, a shadow in the water, the dip of a lily pad followed by the unmistakable *plop* of an animal diving for safety—and you home in on that signal and move after it. He said it was sometimes hard to relearn normal vision after a long session of collecting. He recalled that in the old days when he collected chorus frogs at night, watching for them as his headlamp swept slowly forward in the swamp, he'd often be startled out of his skin when something else—a fish, a snake, a bat—surged into view. "When you really get into it," he said, "the frogs are all you see."

A lot of frogs were tucked back inside the reed line. Most of them hung motionless in the water until we approached, but there were also occasional groups gathered on the muddy shore that would rocket off in all directions as we came near. The day grew hotter and stickier as we worked our way up and down the shoreline. We caught a lot of frogs in the shallows—Erik was amazingly efficient—and Hoppe picked off quite a few that ventured out toward deeper water. In addition to the live animals a fair number of grayish corpses floated belly-up amid the vegetation. Most were tadpoles, but there were also some small fish. The ones that were further gone had begun to dissolve a little and appeared surrounded by halos of decaying flesh. Hoppe collected a sample of the dead animals in a jar for later examination back at the lab in Morris. He much preferred to take freshly dead animals over live ones if he had the option, he explained, even though he expected none of the deformed frogs would survive and that the worst ones would be dead in a few days at most. As bad as it was at CWB, Hoppe said he didn't want to add to the death toll.

After a couple of hours Hoppe asked everyone to report on how many they had. "That's it," he said. "Let's sit down and take a look at them."

Rhonda had set up more lawn chairs in the shade of a large oak on the hillside just up from the lake. Dennis was up, too, and everyone sat down. Rhonda brought out lemonade—it was brutally hot now—and poured big glasses for everyone but Hoppe, who pulled out a can of Diet

Dr. Pepper from a small cooler he'd carried down from the Jeep. We emptied our pillowcases into one of the two deep buckets Hoppe had postioned on either side of his chair. Reaching into the first bucket, Hoppe would snatch up a frog for examination. After he was done it went into the bucket on the other side—and from there, when everything was finished, back into the lake.

It had been clear, as we'd pulled one wriggling frog after another from the black sediments in our nets to sack them up, that many of them were abnormal. But the actual count was astonishing. Altogether, 157 of a total of 202 frogs were deformed. All of them were juvenile mink frogs. The mix of deformities was the usual—skin webbings, split and twisted limbs, extra legs and stumps of extra legs, missing legs and partially missing legs. Hoppe examined the animals one by one, measuring each from snout to vent with a micrometer and calling out their deformities as I recorded the information on data sheets. Hoppe gave the frogs' length, which he measured in millimeters, simply as a number. This was followed by a pig Latin of quick abbreviations for most of the abnormal conditions. For example, a skin webbing—Hoppe's so-called cutaneous fusion—of the left rear leg was recorded as "LR CUT." But even with all the shortcuts I often had to cram to make the descriptions fit into the abnormalities column, writing as small as possible and sometimes carrying the notes out of the allotted space and into the margins and down the side of the page.

About midway through Hoppe's inspection the wind rose and went cold as the sky blackened without warning. Dennis suggested we move up the new pole barn, which was pretty much completed except for the big double doors at either end. As we gathered together the gear and the lawn chairs and scrambled up the hill, enormous droplets began to fall from the speeding clouds and a denser wall of rain from the west swept across the cornfield toward us. We got set up again just inside the open door frame at the east end of the barn, where it took a couple of minutes to adjust our eyes to the darkness. Rain blew into the barn at the other end on a sharp slant, backed by a fizzy, ochre-colored light. We had to almost shout over the hammer of rain falling on the sheet-metal roof—a low, steady roar that was now punctuated by claps of thunder.

"Thirty-six, with a tail of twenty-five," Hoppe yelled out as he took the measurements of the next frog. "RR CUT. LR missing tib-fib! Looks like a stump of a foot attached at the knee." And so on. I wrote it all down as best I could. The descriptions of some of the more convoluted deformities were difficult and, in a few cases, hopeless.

Hoppe's mood seemed to match the storm, though his fury was quieter. He told me he was interested in working on this site more than any of the others because of its terribly high incidence of deformities and because he felt an obligation to try to help the Bocks find out what they were dealing with. The MPCA's efforts thus far had been a disappointment; the agency obviously didn't share his view that the answer to the whole deformities problem was probably right here, in this one spot. Just the week before Hoppe had been interviewed for ABC's *Nightline*. Sitting by his frog tanks at his lab in Morris, Hoppe was a picture of ashen discomfort in front of the camera. Part of it was undoubtedly his basic mistrust of the media—he'd been hounded by reporters for months and had begun screening all his calls before answering the phone—but mainly he just looked worried. Hoppe showed off a couple of deformed leopard frogs—one with an extra forelimb and another with an abnormally pigmented eye, neither of which was an especially common type of deformity—and said he doubted the problem was related to what the animals ate or the ground they lived on. "It looks like it's the water," he said.

Now Hoppe was staring at a third consecutive season of deformed frogs at CWB. He said this year's batch was the most distressing yet, and the many dead frogs floating in the water was an even more ominous sign. In 1996, Hoppe had found the deformities rate among mink frogs there to be about 50 percent. Now more than three-quarters of the frogs were abnormal and another bunch was dead and most likely abnormal as well. Hoppe said things were falling apart here. "This is much worse than last year," he said. "There are more deformities and the deformities are more severe."

Hoppe said he was frustrated by the MPCA's far-flung efforts in the field this season when there was such a dramatically affected site staring everyone in the face that was begging to be intensively studied. "If anybody can figure this thing out anywhere," he said, "it ought to be at this place."

Dennis Bock told me later he was unhappy, too, about the failure of the state and federal agencies working on the problem to undertake more thorough testing of the water at CWB. He wasn't worried for himself, Dennis said, but rather for his future grandchildren. He said he was somewhat encouraged by the fact that his neighbors had lived here much longer than he had and showed no ill effects, at least none that appeared obviously connected to local water quality. "I keep referring back to that as a sign of hope," Dennis said. "But I'm not all that hopeful, actually."

Hoppe visited CWB once or twice a week all through the summer of 1997 and his initial perception that the situation there was worsening seemed to hold true. Two days after the first big collection of mink frogs, Hoppe and the Mottls caught another good-sized sample and got the same result as before. This time—without the storm—they were also able to visit nearby Butternut Lake, the designated control site for CWB.

It's about the same size as the Bocks' lake but deeper and without development anywhere along its swampy, wooded shoreline. Collecting frogs there had proved much harder than at Bocks'. The main point of access—a steep, deeply rutted clay ramp—comes out of the forest at a boggy place on the north end of the lake. Thick islands of floating vegetation hug the shore and make the wading treacherous, as it is hard to tell solid ground from thin patches where a wrong step can plunge you into chest-deep water. In an hour of hard work, Hoppe and the Mottls managed to find only two mink frogs—both adults that appeared to be normal—and a handful of newly metamorphosed American toads that had begun arduously working their way up from the water's edge. Toadlets are extremely small, on the order of a quarter-inch in length, and without caution you are much more likely to step on one than notice it.

Hoppe speculated that the lack of activity at Butternut was probably due to it being deeper and therefore colder than CWB. This would delay the peak of metamorphosis, perhaps signficantly. Amphibians are ecological opportunists, especially in extreme climates like Minnesota's where favorable breeding conditions arrive suddenly and can vanish almost as quickly. Hoppe told me there were three distinct reproductive "strategies" used by the various species in Minnesota. Leopard frogs, wood frogs, spring peepers, and chorus frogs are all early breeders that can tolerate very cold water. By reproducing at the first sign of spring their tadpoles can feed and grow through the first part of summer and reach a good size by the time they metamorphose. Mink frogs and green frogs, the two most similar species in this part of the world, take the opposite approach, breeding later on, after the water has warmed. Their tadpoles then take about a year to develop, overwintering beneath the ice and metamorphosing— again at a healthy size—the following summer. American toads, the most explosive breeders of the lot, reproduce late—their high, trilling song is a major chord in the night music of summer—but lay as many as 8,000 eggs and reach metamorphosis in a matter of just a few weeks. They come out of the water tiny, but in huge numbers.

Hoppe was worried that these timetables were being artificially compromised at CWB. He continued to find hundreds of juvenile mink frogs on each visit well into July, but the leopard frogs were scarce and seemed to have stalled just short of metamorphosis. On August 8, 1997—two years to the day after Cindy Reinitz and her students stumbled onto the deformities at the Ney farm—Hoppe finally began finding leopard frogs dispersing across the Bocks' lawn. It was a cool, beautiful summer day. It took a couple of hours to scrounge up a collection of fifty leopard frogs; Hoppe had hoped for a sample of at least a hundred. Only one of the animals exhibited any obvious deformity—a skin webbing on one hind leg. But among the thirteen straggler mink frogs he caught the same day, seven displayed the usual assortment of limb abnormalities. Hoppe said he believed they would begin finding more leopard frogs as the season progressed, and that if his previous observations held up it was likely that the deformities rate among them would climb as well.

And so it did.

On August 12 leopard frogs were scattered all over the lawn, with a few hanging out near the beach. In an hour and half he'd caught 148 juveniles. Nine were deformed. Just eight mink frogs turned up that day—six of which were deformed. One of these abnormalities was really odd, even by mink frog standards: The frog had an extra hind foot with one long toe, the tip of which was "fused" back to its body, making a weird, closed-loop of strung-out flesh. But Hoppe was more concerned about the number of frogs he was finding. The leopard frog population appeared to be significantly reduced. The year before, he said, they had easily caught upward of 600 leopard frogs at a time in a succession of collections during the metamorphic peak in midsummer. "We'd get tired and quit after taking data on four hundred or so," he told me.

Hoppe was becoming convinced that delayed metamorphosis and the limb deformities at CWB were in some way related. On September 18 he caught 110 leopard frogs. Twenty-five were deformed, notably in the calf region involving the tibula/fibula. Many of the animals were missing most or all of their lower legs, but still had one or more feet with extra toes growing just below the knee.

Delayed metamorphosis. Dead frogs belly-up in the water. Increasing rates of deformities that were growing progressively more severe. Hoppe began to wonder how many more seasons he or anybody else would even have frogs to study at CWB. His fears for the population limited what he felt he could in good conscience do in the way of research. Hoppe did not want to take any animals out of the lake if he could help it, even the deformed ones. He routinely let frogs go after examining them. Sometimes he'd take a few to the lab at Morris, photograph them, and then drive all the way back the next day to let them go, in what was surely a futile effort to help preserve the population. "I know it probably doesn't matter," he told me. "They're not going to make it anyway. They're going to die. They're certainly never going to reproduce. But I just can't keep them."

Hoppe did manage one telling experiment on site. In 1996 he collected a mass of toad eggs and divided them, raising half back at Morris in laboratory water and placing the rest in a cage in the shallows at CWB. The toads raised in the lab were all normal. Of the 57 toads that metamorphosed at the lake, 6 were deformed. All were missing hind limbs or part of hind limbs. This was an admittedly small sample, but the rate and the type of deformity was pretty much consistent with what Hoppe had observed in American toads at CWB. The experiment also suggested that there was indeed something in the CWB water that directly caused the abnormalities. But what was more interesting was the fact that missing limbs had turned up in animals raised *inside a cage*—though it took Hoppe more than a year to realize what it meant. He said he was almost embarrassed he hadn't seen it right away when he told me about the results months later, but the experiment also showed that the missing legs couldn't have been caused by predators. "At least not unless Stanley can tell me what predator can open up a cage, climb in and eat a leg off a toad, and then climb out and close the cage again on its way out," Hoppe said.

Hoppe had gotten over the shock of finding so many deformities in 1996, but his personal discomfort with the situation in Minnesota was steadily growing. In early August he'd gotten a call at his home from a resort owner on Lake Mary in Alexandria, a vacation hub in west-central Minnesota. It seemed that a kid fishing off the dock had discovered a leopard frog with extra hind legs. Hoppe went over to investigate and collected 107 young-of-the-year leopard frogs. Sure enough, eleven of them were deformed—not the catastrophic rate he was seeing at CWB, but still way above normal. Hoppe took some of the abnormal animals home to photograph. His phone began ringing crazily. The resort owner—apparently believing it would somehow be good for business—had also alerted the local media. Hoppe ducked the reporters' questions. When he took the frogs back later that day to let them go where he'd found them the place was staked out by a television crew. Hoppe drove home and then went back again much later that night and released the frogs under cover of darkness. When he told the MPCA about the discovery, Judy Helgen asked if they could collect some frogs from the site to send to Madison for evaluation. Hoppe, fearing that a media circus might ensue on such a densely populated lake, told her no.

That same summer I went to see Andy Blaustein in Corvalis and he took me with his team of graduate students up into the Cascades to visit a couple of his study sites. It was beautiful country—steep mountains and towering Douglas firs giving onto stunning vistas of yet more steep mountains and towering Douglas firs. Blaustein likes to say that pristine locations no longer exist anywhere, that no habitat has totally escaped the influence of human progress. But as the Oregon State University van climbed the switchbacks through the shadow and light of the high Cascades it felt as if there might yet be a few places that qualify. Nature was all around us in full force, in the trees and the rocks and the shimmery sky overhead.

Blaustein hadn't started out studying amphibians. After growing up in Brooklyn and going to college on Long island he'd gone west to graduate school, first at the University of Nevada and later on to the University of California Santa Barbara, where he studied animal behavior. "I am a card-carrying behaviorist," he told me more than once. His interests were not toxicology nor even primarily amphibians, even though his work in recent years had centered on frogs and salamanders. Blaustein's early fieldwork was on the population dynamics of small desert rodents. At Oregon State he began working on kinship recognition and tadpoles had proved to be a good laboratory organism for those experiments—at least until Blaustein started having trouble finding frogs in the nearby mountain wetlands in the mid-1980s and his attention was suddenly diverted. When I asked him about those early disappearances he answered sharply. It hadn't taken him long at all to formulate an idea of what was happening. It was, he said, really pretty obvious after they'd sorted through the data. At first they noticed that Cascades frogs (*Rana cascade*) were becoming scarce. Then they began finding dead egg masses laid by western toads (*Bufo boreas*). Even though the toads themselves were still all over the place, their eggs were dying at alarming rates. "Really massive numbers," Blaustein said. "It was ninety-five percent in some places, a hundred percent in some others. The historical records we have say they don't usually do that. They usually have less than ten percent mortality rates in the egg stage."

"But when you brought these eggs into the lab did they hatch normally?" I asked.

"Correct," he said. "Yep. And in the same water, too. We bring in the same lake water they're breeding in. Everything is the same except that they're not outside. We couldn't find anything wrong with the water, no pollutants. So we began to wonder about the possibility of UV radiation."

Blaustein's work had long since established a UV connection in the decline of several amphibian species in the Cascade Range. He'd shown that species with a weak photolyase response were the same animals disappearing from the mountains. Photolyase is the enzyme that promotes repair in UV-damaged DNA.

Now he was trying to assess specific effects based on careful direct monitoring of ultraviolet exposures in the field. The experiments involved rearing tadpoles and also salamander larvae in enclosures in their native wetlands. These enclosures—square boxes about three feet on each side that sat right down in the water where the animals normally lay their eggs and hatch out—were covered with filters that either transmitted or removed UV-B from the light spectrum. Blaustein's team had to monitor the actual amount of ambient UV falling on these sites several times a week. For more than a decade now, he'd been working on the assumption, widely shared among atmospheric researchers, that Earth's depleted ozone layer was allowing significantly higher levels of the dangerous UV-B portion of the spectrum to reach the surface of the planet. But the problem of verifying this suspected increase was not a simple one.

"Every week I get the ozone readings from Canada and how much it is deviating below the norm," Blaustein said.

"What's normal?" I asked

"I don't know. Nobody does." Blaustein said the challenge was to set a benchmark against which changes in UV penetration of the atmosphere could be tracked.

"If I look, say, at the numbers for Vancouver, which is not very far away from us, and I see that the ozone layer is running 11.3 percent below normal for that region, you'd think I could simply calculate out the increase in the UV associated with that.

"But I can't. You know why? 'Cause it's changing all the time! It changes every second. A cloud comes by and it changes. Something else—

the water's not as clear, whatever, it just changes. It's fluctuating all the time, so you can't do anything with that approach. You can't just say UV is up this much or that much. But what we find using these enclosures is that you can measure some very important differences in what happens to organisms that are living under filters that block UV. Some very nice data."

I found Blaustein to be a different sort of biologist. He was direct, snapping off answers and opinions at will. He always seemed to know what he thought, even on the kinds of questions other scientists mulled over before answering. Eventually it became clear that Blaustein's brash style was just the way he was, and that for him all questions about ecological interactions led to a single basic conclusion: Things are never as simple as you think they are or might wish them to be. In essence, this was the answer he gave to just about everything I asked him.

Blaustein had found that UV could cause frog eggs to die. Also that it could cause retinal eye damage, skin lesions, minor digit deformities, spinal curvatures, and a weird, balloonlike bloating or edema in the abdomen. He'd also discovered that UV could weaken frogs, presumably by stressing their immune systems, so they were readily infected with a fungus called *Saprolegnia ferax*. This was interesting because *Saprolegnia* is normally a fish fungus. Yet Blaustein found it in frogs—even in wetlands that didn't have any fish in them—and eventually determined that it could be transmitted from fish to frog and subsequently from frog to frog. Even insects appeared capable of transmitting the fungus to amphibian ponds. The world was a busy place. "We think the UV compromises the frog's defenses to this kind of fungal infection, but also to disease in general," Blaustein said.

Blaustein and his team checked their enclosures at a site he called Small Lake, which was a shallow, temporary wetland that dried out as the summer progressed. Already it was only slightly more than ankle deep in the area along the reed line where the experiments were set up. This kind of wetland is often preferred by midwestern frogs—shallow ponds are usually without fish predators—and Blaustein said that was true here, too. But he assured me the tadpoles would be long gone before the pond dried up. "They have to get out of the water before the end of August because that's when it will start to snow," he said. "Some species will lay their eggs in the shallowest possible water or even slightly out of the water to put more heat on them and speed up the processs."

I said I thought it was surprising, given this kind of adaptive behavior, that some species had a robust photolyase response that allowed them to repair UV damage to their DNA, while others did not.

"I know what you're thinking," Blaustein said. We had just finished up at Small Lake and were standing around back at the van. Everyone was smoking cigars—a post-fieldwork ritual on Blaustein's teams. In his trademark UV-fighting dark glasses and long sleeves, Blaustein went on, wreathed in smoke.

"You're wondering why—since they all live in the same ponds—why aren't they all in the same situation with respect to UV? Why don't they all have the same protection with photolyase?"

I said that was exactly what I was wondering.

"Good!" Blaustein exclaimed. "That's the question. The answer is because things are complex. There might be historical reasons. There are different microhabitats within the same pond. And there are different breeding strategies and time periods. Lots of stuff. But in evolutionary time there was undoubtedly a strong selection pressure for some of these animals to lay their eggs in shallow water, where it was warmest, and to get out of the pond fast. Now that we've screwed up the ozone, maybe that strategy is killing them."

On our way back to Corvalis, Blaustein told me he was still appalled by the deformities he'd seen photos of at the Shenandoah meeting. The van swayed gently as we bent through the turns, back down into the Willamette River Valley. Blaustein said he thought Stan Sessions probably had it right, though. He hadn't caught any deformed frogs himself, but a few had been discovered in Oregon and turned over to him. They were *Hyla regilla*, the Pacific tree frog. "I just figured it was parasites," Blaustein said. He'd sent the animals out to Sessions, and, sure enough, Sessions had found them loaded with trematode cysts.

"What were the deformities?" I asked.

"Extra legs," Blaustein said. "Extra hind legs."

I said that was interesting, because most of the deformities that were being reported now weren't extra legs at all. It was other things. Skin webbings. Missing legs.

Blaustein sat up a little and turned toward me.

"I have a problem with that," he said.

I told him this was the big question now. Maybe parasites could explain extra legs, but they didn't seem to explain missing legs.

"No," Blaustein said, "but a predator could."

I didn't think I'd quite heard right.

"But if it's predators," I said, "why don't we see these deformities everywhere there are frogs?"

"What?" Blaustein was suddenly very focused.

"Missing limbs. Why don't predators bite them off all the time?"

"Are you sure parasites can't do something like this, just leave a stump or something?" Blaustein said.

I said that I recalled Sessions saying they might, but that I was certain he would still attribute most missing limbs to predation.

"That's what he says? OK. It's his theory. So then he has to say everything else is predation."

"Yes."

"I don't know about that." Blaustein looked out at the countryside rolling by. We were back in the flatlands now. Whitish-purple blackberry bushes choked the ditch alongside the roadway.

"I have to tell you this," he said finally. "Almost never do I see missing legs."

We rode along quietly for a minute.

"I think parasites are involved," Blaustein said. "I'm convinced by that little experiment Stan did with the beads, because I'm an experimental guy. I'm also convinced a lot of these frogs have parasites in them. But do I think parasites are gonna explain every single case? No."

"We've Found Something in the Water"

As THE AUTUMN OF 1997 APPROACHED, THE DEFORMED FROG investigation entered a new phase. After almost two years, the inquiry had grown large but disjointed, and the sequence of events—bafflement followed by a largely ineffective search for common factors across a vast geographic range of outbreaks—didn't seem to be leading anywhere. Inevitably, the outbreaks lost some of their shock value and investigators argued over the significance of what was being seen. The main difficulty—distinguishing deformities that had natural origins from those that might have unnatural causes—was all but ignored. There remained a presumption that these two separate categories existed, but somehow the word *deformities* became accepted as a universal term that meant the same thing everywhere.

As doubts stole into the investigation, some researchers, such as Dorothy Bowers and Joe Tietge, started to wonder if they were even looking at an actual problem. Since none of the research approaches overlapped—everyone was doing something different—the data acquired did not mesh, either. There were striking differences among the sites investigated, among the deformities detected, and among the approaches taken, and as long as nobody could compare their results with anybody else's, the picture of the deformities problem became more detailed but no clearer. Like a group of people describing the countryside surrounding a large house as each looks out a different window, the scientists working on the deformities depicted the terrain in vaguely similar ways, but the particulars varied considerably.

The impasse ended suddenly, when the investigation came explosively unblocked in a succession of surprising and provocative findings.

In July of 1997, Dennis and Rhonda Bock noticed that researchers from the EPA and the MPCA had begun taking large amounts of water from their lake as well as from some tubes that had been installed near the shoreline to collect shallow groundwater samples. Eventually they took some water from the house as well.

"I remember that Judy Helgen came one day by herself and took a lot of water, a lot of big jugs of water," Dennis told me.

"Yes," Rhonda added, "and she said she was going to run it up to Duluth. She must have had fifty jugs. They took water out of the outside faucet, too."

At the EPA lab in Duluth, surface water from the Bocks' lake was being evaluated in a battery of standard toxicity tests on several larval organisms commonly used for the purpose, including fathead minnows, a kind of water flea called daphnia, and freshwater shrimp. Joe Tietge told me the results of these tests had been negative—except for a single sample in one of the bioassays using shrimp. "We got 100 percent toxity in that one," he said. In other words, all of the test animals died. Preliminary fractionation of the water sample suggested, though not conclusively, that the lethal agent was not an organic compound, meaning probably not a synthetic chemical.

Tietge said the EPA considered this result "marginal" in its significance, and the feeling at the lab was that freshwater shrimp might just be unusually sensitive to some imbalance in the normal constituents of the water. But Helgen was sufficiently alarmed by the findings that she insisted Jim Burkhart accelerate his scheduled testing of Minnesota water samples with FETAX, which he did. FETAX, the 96-hour assay using embryos of the African clawed frog that Burkhart had planned to use to isolate biologically active fractions from water samples had a long and generally reliable track record in detecting certain known toxicants, many of which left telltale "signatures" in the way they affected the embryos. But now FETAX was being pressed into service as a more general screening assay in a rush to answer the question that had haunted researchers from the start. Was there something poisonous in the waters of Minnesota? In September, the first hint of what had been found that summer reached the Bocks via David Hoppe.

Hoppe and Erik Mottl were waiting for me at the Bocks' on a bright, bejeweled day in mid-September. Hoppe had told me earlier in the week that in addition to a routine collection at CWB they planned to spend part of the day surveying several of David Merrell's original leopard frog collection sites in the area to determine how those populations were faring and whether there were deformities present now that hadn't been seen historically. Hoppe's Jeep was crammed with gear when I arrived, so I offered to follow them. But Hoppe shook his head and shoved a couple of nets into the back.

"Get in," he said. "I've got news."

We drove off down the Bocks' driveway dustily, the stippled surface of CWB retreating behind a wall of autumn leaves in the rearview mirror. Hoppe seemed agitated. He said he'd gotten a call from Judy Helgen earlier that morning. She'd heard from Jim Burkhart, who'd received a batch of FETAX results back from the Stover Group in Oklahoma, which was under contract to the NIEHS on the project. They were still working to process samples taken at a number of sites in Minnesota, including water collected from more than a dozen private wells. But the initial assay results of surface water, groundwater, and bottom sediments from sites with deformities were positive, producing both high rates of mortality and malformations in the lab embryos.

The samples were toxic. The samples were teratogenic. They produced developmental abnormalities.

"Judy wants me to let Dennis and Rhonda know that they're working on their well water now," Hoppe said forlornly.

The implications of these findings were uncertain. Hoppe didn't know at this point how strongly the results correlated with deformities confirmed in the field. Nor did he—or anyone else—know for sure what these results portended for other species, including humans. But Judy Helgen had already decided that the MPCA would take no chances. Her promise to "tell everyone" when they found something was about to be fulfilled.

"I think she's planning a press conference of some kind," said Hoppe. He slammed the Jeep to a halt in the parking lot of a boat ramp on the lake we'd arrived at, then turned around to look at me. I was trying to recall how many times he told me he had begged Helgen to test the Bocks' well water over the past two years.

"What I want to know," he said almost inaudibly, "is what in the hell I'm supposed to tell these people."

The Bocks took Dave Hoppe's halting explanation about the FETAX results in stride. Hoppe cautioned them against assuming the worst—whatever that might be. Dennis shrugged and said something to the effect that "it's about time" they were checking on the well and he hoped the MPCA would keep them informed of what was going on. Both he and Rhonda had reservations about the MPCA. Dennis told me they liked Judy Helgen well enough, and they were especially impressed with the super-diligent Dorothy Bowers. But they admitted they were disappointed at the slow progress of the investigation. Dennis said he didn't see why the same tests couldn't have been done at least a year earlier. "For an agency that is supposed to protect the people of this state from pollution, they were very slow getting geared up," he said. "On the whole, I haven't been all that impressed with the MPCA."

I didn't go to Judy Helgen's press conference in person, but I didn't have to. The whole world was invited, thanks to *Nightline*, which timed its second program on the deformities problem to broadcast the MPCA's announcement of the FETAX findings on September 30, 1997. I watched it on TV, at the end of a two-day scientific retreat at a waterfront resort in Southern California, where Dave Gardiner and Sue Bryant had quietly convened a brainstorming session with several colleagues on what might be causing the deformities. Ever since the Shenandoah meeting the previous spring, Gardiner had wanted to be directly involved in the investigation. In August, he told me that he and Ken Muneoka had left Shenandoah extremely frustrated at the direction of the ongoing research. "I feel very strongly and Ken feels very strongly," he said, "that you're not going to get at this in a meaningful way by simply doing field studies."

Toward the end of summer, Gardiner and Bryant decided to get together with Muneoka and another old friend, Bruce Blumberg, from the Salk Institute in La Jolla, a private research facility that is home to the world's most prominent hormone receptor lab. They hustled up a $5,000 grant from UC Irvine to rent a pair of adjoining condos on a canal in Oceanside where they wouldn't be disturbed, and to later pay for some sort of pilot study if one could be devised. Also at the meeting were Stephane Roy, a postdoc, and Mark Carlson, a graduate student, both from Bryant's limb lab, plus Val Lance, a wildlife endocrinologist from the San Diego Zoo, and Tyrone Hayes, a herpetologist from Berkeley. The mood was casual, but there were orderly presentations both days, interspersed with long periods of freewheeling analysis. In the evenings, Gardiner held forth while making sure that the group embarked on prolonged happy hours that went past midnight.

The technical discussions focused on two essential considerations. What would cause a frog to grow malformed limbs and how would you detect such a causative agent in the environment? Ken Muneoka, in a talk similar to the ones he'd delivered in Duluth and at Shenandoah, again outlined the basic morphological and molecular sequences that initiate limb outgrowth and guide proper pattern formation—subjects on which most of those present were more than conversant. Muneoka emphasized that in experimental work, especially on mice, exposure to a teratogen had to occur very early in limb development to create extra legs or digits. "We need to focus on what's happening in the limb field," Muneoka said, meaning the region of cells on the embryonic flank that are fated to form a limb. If the limb in question were an extra one, then something had to establish a false or enlarged limb field from which it could grow.

This was the place, everyone agreed, to begin trying to understand how an abnormal leg would develop. It seemed all but self-evident that leg malformations could *only* be caused by something that disrupted the biochemical signaling necessary for the correct initiation of the limbs. But what was it? Everyone also agreed that many of the deformed frogs that had been widely seen and photographed—mainly the dramatic, extra-legged ones—exhibited limb abnormalities that resembled the kinds of malformations that had been experimentally induced via exposure to retinoids. And since retinoic acid and its cousins were hormonal in nature, it followed that the deformities could reasonably be hypothesized as the result of a kind of endocrine disruption, either by an actual retinoid or by a mimic. Dave Hoppe had sent out a couple dozen mink frog specimens from CWB, which Dave Gardiner had cleared and stained and which everyone spent some time gawking at under a microscope. Earlier that summer, on the first *Nightline* program, Gardiner had said unequivocally that he could create deformities in laboratory animals that were similar to these kinds of abnormalities. And he'd said more.

"You know those frogs?" Gardiner had said, in reference to the reports coming out of Minnesota. "Well, pretty soon you'll have a person who looks that way."

This was an admittedly sensational claim, but one that accurately reflected the real concerns of the biologists in the room at Oceanside—who as a group worried that limb development mechanisms are similar among vertebrates and who were also used to thinking of frogs as a model species that illustrated what could occur in higher organisms like humans. Seeing abnormal limbs in frogs was another day at the office for them. As Sue Bryant told me later on, forcing limbs to grow incorrectly—as in grafting experiments—was a common feature of all the research on limb development. It was even possible, Bryant maintained, that what was known about limb development pertained more directly to abnormal limbs than to normal ones. "You could argue that development undisturbed is a different process than development disturbed," Bryant said. "So that means that we really don't know how limbs develop. And there are some results that suggest that's true." After years of making "deformed" amphibians in their labs in order to understand the process of limb development, it seemed now that—just as Gardiner had said at Shenandoah—they were faced with the results of a similar experiment taking place in nature.

The meeting was not a blind search for an approach. In fact, Gardiner, Muneoka, and Blumberg had already formulated a research plan that they'd laid out for Joe Tietge in an effort to join his project. Those discussions had come to an unhappy end when Tietge explained that the EPA didn't have any money to contract with them—a fact that the California researchers said only came out after they'd already explained how to do the experiments. Gardiner, Muneoka, and Blumberg were now soured on collaborating with anyone else on the deformities investigation.

"Fool me once," Blumberg said.

Blumberg, a burly butcher's son from New Jersey, had interests in both developmental and molecular biology. I found him irrepressibly curious about every subject that came up—not to mention opinionated, rough, combative, quick with an obscene joke, and very, very intelligent. He was great. Blumberg's main area of study was a family of highly specialized receptor proteins that lie inside the cell, immediately connected to DNA in the nucleus. These so-called nuclear hormone receptors initiate gene expression when they bind with the hormone for which they are dedicated, and it's this direct linkage to the DNA that makes many steroid or vitamin-derived hormones so powerful. The binding and activation process is an interlocking sequence—like a key turning a lock that opens a door. The right key always opens the door, and the right hormone, or "ligand," always activates gene expression. The result of gene expression depends on the hormone and the type of cell involved, but hormones and genes drive all hereditary and developmental features of an organism, and also perform "housekeeping" functions—such as regulating metabolism—on a continuing basis throughout life. Hormones began to be accurately characterized in the 1950s; in the 1970s the first hormone receptors were discovered. Many are now known, as are their ligands. Work continues to identify the ligands for "orphan" receptors—receptors that have been found but whose function is unknown. Blumberg was among the world's leading experts on nuclear hormone receptors and had personally already discovered the ligands for two of the orphan receptors.

This work had a number of potential practical benefits. Receptor studies are anticipated to lead to medical breakthroughs in the future as gene therapies are perfected. More immediately, identifying receptors and their ligands can help scientists understand the process of endocrine disruption. In a case where suspected endocrine disruption involves a known hormone and a known receptor, it's possible to do an assay that may identify a compound that is mimicking a hormone and binding to its receptor.

This was the working hypothesis of the Oceanside group, and it was Blumberg who could test it, because the receptors for retinoic acid and its analogues were known. They could be cloned and used to search for an active ligand in an environmental sample. Blumberg had helped pioneer these techniques, including a fully automated robotic assay system he'd developed that was both sensitive and fast. In other words, if there was a retinoid or a retinoid mimic in Minnesota water, Blumberg could quickly test for it and, with a little more work, identify it.

"We're not saying that we know these deformities are being caused by retinoids," Gardiner told me once again. "We're saying that's what it looks like and that ought to be the first thing we look for if we can. Well, Bruce can."

The discussion of multilegged frogs seemed to leave some important questions unanswered, and I brought up a few of these to the Oceanside group.

Why was it, for example, that so many deformities in the field were unilateral? If the malformation was caused by some chemical the frog was swimming in why would it affect only one side of the animal? Gardiner offered to think about it. Ditto my inquiry about cutaneous fusions and missing legs—morphologies that occurred at many more sites than did extra legs. Bryant seemed surprised by this information.

"Skin webbings are more common, you say?" she said.

I said that was the case in some places, but not everywhere. The real point was that it didn't appear that all deformities could derive from a singular cause—since they didn't appear to be a single phenomenon. I said I didn't think there was any general phenotype that was representative of everything that was being seen, but that missing legs seemed to be the dominant problem and that cutaneous fusions seemed to be a category of deformity that hadn't received enough attention.

"Well, all I can say at this point is that skin is very important," Bryant said. "It takes a competent epidermis to grow a leg."

This was one of the few comments Bryant made directly on the deformities all through the meeting—during which she mainly listened. "This is all news to me," she said. "I mean, I've talked to Dave and Ken about these deformities, but they're very puzzling."

I'd gone to California hoping for more insight than this from the woman who seemed to be a sort of fairy godmother to the whole frog investigation. Both of the main hypotheses about the causes of the outbreaks—chemicals and parasites—had origins in her lab. Implicitly, Bryant concurred that the retinoid hypothesis should be investigated at once. And she offered definite, fixed perspectives on limb development. But she also seemed open to new ways of looking at the deformities problem that might emerge as more was learned. Bryant was originally from Sheffield, England—she'd first come to the United States as a postdoctoral fellow—and her precise British accent gave her courteous statements a certain gravity. On several occasions she scolded people for making fun of contrary viewpoints—especially Stan Sessions's. It was easy to see why so many people looked up to her.

Some time after the meeting, Bryant told me that she kept a "special compartment" in her brain that was reserved for information she couldn't make sense of right away—a kind of holding cell for data that needed to be taken out and examined more closely at a later date. The upshot was that the person who arguably knew more about limb development than anyone else involved in the investigation was the person most cautious in her speculation as to what was happening.

Nightline came on a few hours after the close of the second day of discussions at Oceanside. Everybody waited around anxiously on the patio or walked down to the canal to watch the lobster divers swimming at night with their lanterns. When the program finally began, MPCA commissioner Peder Larson appeared onscreen, summing up what *Nightline* host Ted Koppel called the agency's "dramatic announcement."

"We're here because we've found there's something in the water," said Larson.

The press conference was a joint effort by the MPCA and the NIEHS. Larson was seated at a long table and was flanked by Judy Helgen, Jim Burkhart, and George Lucier. Lucier summarized the findings from the FETAX studies.

"We found that water from sites where malformed frogs have been reported was very potent in deforming frogs in this laboratory experiment."

The MPCA said the assay had been completed on pond water, groundwater, sediments, and now well water from Minnesota hot spots and control sites. The results were positive in samples from sites with deformities and negative for the controls. In light of the well-water findings the agency was taking the precautionary measure of distributing bottled water to families whose wells had tested positive in FETAX. In an oddly self-nullifying attempt to soften this news, Larson warned that what the agency really meant was that it didn't know what it meant. "We have no reason to believe or not to believe that anything you're hearing about today is transferrable into any human impacts."

Over coffee and doughnuts poolside the next morning, the Oceanside group chatted about this new development and the weird half-alarm being sounded by the Minnesota researchers. In general, everyone seemed to think that *Xenopus laevis*, the African clawed frog, was a standard lab animal so FETAX had to be presumed to have validity. "Heck," said Ken Muneoka, scratching his head, "they must have a genuine concern." Muneoka found it unlikely that the MPCA or the NIEHS would have volunteered the FETAX results unless they were trying to avert a health disaster or a panic, or both, since scientists generally keep preliminary findings to themselves. "I mean it's gonna be really hard for them to get this stuff published now that they've gone public with it."

This was a generous interpretation. It was the last one I ever heard anyone offer on the subject.

Since none of the FETAX results hinted at what it was in the water that caused death and deformities in the assay the announcement in a way only seemed to reiterate the obvious. Hadn't nature already done the same experiment? What, really, was the difference between a wild frog with a deformity and a lab frog reared in the same water that developed a deformity?

As it turned out, quite a lot. Within a month the MPCA and NIEHS would endure a startling reversal of fortune, as experts lined up to attack the FETAX results. Within four months, they'd end up taking it all back—though not with a press conference the second time around. In the meantime, the MPCA's eagerness to announce that they were taking precautionary actions to ensure safe drinking water supplies to people living near deformities sites was not matched by their success in actually implementing the plan. When I spoke with the Bocks nearly two weeks after the agency's press conference they had yet to receive any bottled water, despite several calls they made to St. Paul to find out what was going on. Dennis was eventually told that the delay was due to Judy Helgen's concerns about maintaining the secrecy of the CWB study site. Rather than send out water directly from a supplier in Brainerd, a complex middleman arrangement had been devised, apparently involving another water distributor in the town of Wilmar, nearly one hundred miles away. Rhonda was under the impression that the final delivery truck would receive directions to the house but no address or phone number.

"I don't want to seem ungrateful that they are making this water available to us," Dennis said. "but are they really making it available to us when we don't have it? If it's important for us to be drinking bottled water we ought to have it in the house."

In the end, the Bocks got their bottled water. They even drank a little of it. But not for long. Rhonda told me the whole family gave it up pretty quickly.

"It tasted kind of plastic," she said.

The infighting that erupted in the weeks following the FETAX announcement was not entirely unanticipated. Two researchers in particular—Martin Ouellet and Stan Sessions—felt all along that the U.S. deformed frog inquiry figured to be a bumpy ride because they believed, for different reasons, that it was on the wrong track to begin with. And both had warned me as much.

Oneonta is a sweetly autumnal college town on the Susquehanna River in south-central New York State. The streets downtown are lined with cozy shops and restaurants, bookstores, a fly-fishing outfitter. Hartwick College, situated on a hilltop commanding lovely views of the town and the river valley, is the epitome of a small private liberal arts college—a compact, refined world unto itself. Stan Sessions, who before coming here had been shortlisted for a faculty position at Harvard that he didn't get, told me that part of the appeal Hartwick held for him was that it made him a big fish in a small pond. The college, he said, was "delighted" with the attention it received as a result of his involvement in a high-profile scientific investigation. At the same time, there weren't a lot of the pressures that college professors endure at larger universities when it came to publishing and landing research grants. Sessions said he was quite content in this more relaxed atmosphere, which he equated with absolute academic freedom. He said he was an incorrigible "putterer" and at Hartwick he could do pretty much as he pleased. Plus, he liked teaching, which was a primary focus at the college.

Sessions had come here after his stint at UC Irvine, and he remembered his four years as a postdoctoral fellow in Sue Bryant's lab fondly. But he told me that in some ways he felt he hadn't completely fit in back there, having never really penetrated an inner circle that consisted of Bryant and two other postdocs: Bryant's husband, Dave Gardiner, and her star pupil, Ken Muneoka. "It was very exciting," Sessions said. "The developmental biology center at Irvine is one of the top two or three development labs in the world. Everybody who was anybody came through there. We didn't all get along all the time, but it was a very dynamic place."

Sessions said he'd been the odd man out for several reasons. He'd gone to Irvine in 1985 as an evolutionary biologist. This, he said, made him out as something of a "weirdo" to the developmental types. At Berkeley he'd gotten his Ph.D. under the eminent herpetologist David Wake, who encouraged Sessions's interest in *Plethodontid* salamanders. Sessions was fascinated with this group of amphibians, native to North and South America, because of their unusually large genome size. Contrary to what you would expect, the size of an organism's genome doesn't necessarily correspond to its complexity. A frog cell contains more DNA by weight than does a human cell. Salamanders have still more DNA. Sessions found that larger genomes make for larger cell sizes, and that larger cell sizes in turn mean each cell in the organism has to do a greater portion of the work of development. It also means that the organism develops more *slowly*, and Sessions had learned how to use a regeneration assay to measure the effects of genome size on the rate of development.

Bryant, an internationally famous pioneer in the field of limb research, had married the processes of development and regeneration to one another by proposing that the fundamental mechanisms governing both were the same—an argument that was strongly supported by experimental grafting results, notably in Bryant's extensive studies of growth and pattern formation in salamander limbs. Like Sessions, who had his own way of doing things, Bryant was also something of an iconoclast. For years she had resisted the widely accepted theory that cell differentiation and pattern formation in the vertebrate limb were determined by some mysterious chemical substance that diffused from an unknown source near the base of the limb. The so-called diffusable morphogen was thought of as a master control compound that spread through the developing tissues of the limb in concentrations that steadily decreased at greater distances from the source. This chemical gradient supposedly gave cells in the limb their identities. When the involvement of retinoic acid in limb development was discovered, it was generally thought that it was the diffusable morphogen.

Sue Bryant was the only prominent limb expert in the world who rejected the idea of a diffusable morphogen. She was convinced the developing tissues in the limb largely determine their own arrangement via localized, cell-to-cell communications—and that the intercalation that took place when cells were experimentally rearranged proved this. In Bryant's version of the halftime show—which she clung to stubbornly for many years in the face of strong criticism from her colleagues—the fans in the stands work together to create a coherent image. Today, the morphogen question has not yet been finally settled, but the increasingly sophisticated molecular studies of the past decade have shown that a number of specific genes and growth factors are involved in limb patterning, essentially confirming Bryant's views. As one biologist recently put it, cell arrangement in the developing limb is "more of a democracy than a dictatorship."

Bryant's lab was just the sort of place where someone with Sessions's background and temperament would be expected to thrive, and for the most part he did. "Sue was a kind of hero to me," Sessions said. "I think she was interested in having me in her lab because she wanted to expand in the direction of evolutionary biology."

It was a fun place, at times. "We did some *weird* experiments," Sessions recalled. "In one, we tried to find out if salamander and frog cells could talk to each other. And they did. When we did the grafts, the tissues worked together. A picture of it got on the cover of *Developmental Biology*. It's this beautiful amphibian hand that's half salamander and half frog. The cells interact with each other. They know what to do. It was a very weird and very dramatic result."

But Sessions said his own interactions with other people in the lab were not always so accommodating as the strange flesh he worked on. I asked him if he was ever at cross-purposes with the people there.

"Sometimes," he said, "sometimes not. I felt I was part of the family. We had differences and stuff. I had the hardest time getting along with Dave. But I think everybody else felt the same way. I mean, I wasn't alone on that. Dave's a tough guy to get along with. He's very emotional. And he's the boss's spouse."

Gardiner later confirmed for me the occasionally strained nature of his relationship with Sessions: "I remember one time when Stan took exception to something I'd said to one of our graduate students. I guess he thought I'd been too harsh in the way I'd dealt with some problem. Anyway, our discussion got kind of heated, so we decided to take it outside. We started out on this long walk, yelling and arguing, and after a while we were way off campus and out in some field, where I guess we were going to fight or whatever."

The Irvine campus is on high ground surrouned by a vast marsh. I tried to picture Gardiner and Sessions, both of whom are tall and lanky, glowering at one another as they strode off toward the savannah.

"What happened?" I asked.

"Well," Gardiner said, laughing, "just as we're about to start duking it out we got attacked by this huge swarm of bees. I guess we wandered too close to them or something. They were just suddenly all over us and we took off running like crazy. By the time we got away from the bees we were out of breath and laughing too hard to fight. So we just went on back to the lab and went back to work."

When I met with Sessions during summer break in his classroom, which was also his lab, he did seem at home. His puttering was much in evidence. There were several large, rectangular plastic containers filled with water and hundreds of lively looking tadpoles giving off a good, swampy stink in one part of the room. Most of them, Sessions said, were *Rana sylvatica*, the wood frog. On another table was an assortment of covered glass dishes, each of which held a cleared-and-stained specimen from Aptos. They looked even prettier than they had in the pictures. We sat down next to two large microscopes. One had a video camera attached to it, and this was connected to a television. On the chalkboard nearby was a handwritten reminder about sorting tadpoles into various limb stages and exposing them to trematode cercariae, over which Sessions had written the word *Save*. I asked Sessions if he considered himself a herpetologist.

"I would like to," he said. "But I'm not a very good herpetologist. I'm too much of a lab rat."

Ironically, it was Dave Gardiner who had suggested the bead experiment to Sessions while he was struggling to understand what had happened at Aptos. When Sessions got his first batch of deformed frogs from Aptos he was amazed. He asked for more and received hundreds just like them. They arrived alive, but Sessions quickly found that he couldn't keep them that way and had to kill them. "That was always our motto at Berkeley," he said. "Kill it before it dies. The problem with these frogs was that they would *drown*. I would keep them in a pan with paper towels and just a little puddle of water to keep them moist. But they kept flipping over on their backs and they couldn't right themselves. So when I saw that they were dying I killed them and preserved them."

Sessions cleared and stained the frogs, which were all Pacific tree frogs, all with hind limb abnormalities, mostly extra legs. The frogs were quite small, so Sessions dispensed with the usual step of skinning them first. The rest was standard procedure. Clearing and staining was an everyday procedure in the limb lab. It was an effective way to see what the internal bone structures of an animal looked like without dissecting it. When Sessions examined the cleared specimens, he was flummoxed. "I didn't know what the hell was going on," he told me. Chemical analysis of the ponds at Aptos came back negative for heavy metals and other probable contaminants, so Sessions kept looking at his cleared-and-stained animals, wondering what to do with them, hoping something would eventually jump out at him.

Something finally did.

"I started noticing these little *things*," he said. Sessions hadn't at first paid any attention to the small, dark spots that appeared clustered near the bases of the hind limbs. But when he finally looked at them under a high-power microscope he saw little mouthlike structures he believed were "suckers." Sessions recalled from his course work in invertebrate zoology that certain flatworm parasites had such orifices. "Once I saw the suckers," he said, "I knew pretty quickly that these were trematodes."

I asked him if the host organism made the "capusle" into which the parasites were tightly curled.

"No," he said. "The worm makes its own container, its own capsule."

"Like a cocoon?"

"Yeah, sort of like that."

"And these cysts were concentrated around the juncture of the limb and the body?" I asked.

"Yes."

Sessions again outlined the basic theory for me. A definitive host—a larger carnivore such as a garter snake or possibly a bird—harbors the mature trematode in its gut. These reproduce and the eggs are excreted in feces dropped in wetlands and transmitted to aquatic snails. With some trematodes, the eggs hatch in the water and a middle stage of the parasite infects the snails. In other species, the snail picks up the egg and it hatches inside the snail into a form that contains many smaller offspring. Depending on the species, there can be several of these intermediate reproductive steps, each adding to the number of parasites, so that in the end a single egg may give rise to as many as 250,000 cercariae that the snail then releases or "sheds" into the water. "Just imagine how many come out if the snail picks up ten eggs," said Sessions.

Sessions said that once he noticed the encysted parasites near the abnormal limbs the rest of the story was easy to deduce. If these metacercarial cysts caused the limb deformities, then the infected frogs were more likely to be eaten by predators that were the definitive hosts—and the circle was closed. Those parasites with a propensity to cause limb abnormalities in their intermediate hosts would, over time, come to dominate the gene pool. In a flash, everything Sessions was interested in came into play. Development. Limb pattern determination. Evolution.

"It just fell into place," he said. "As soon as I saw a trematode, it just all came into focus."

"Did you have any understanding of the mechanism at that point?" I asked. "Did you assume it could be just the physical intrusiveness of the parasites, or did you think they might have some chemical property, something retinoidlike?"

"All of the above," Sessions answered. "We had this aggressive, nasty little worm going into the limb buds of the frogs. *Something* had to happen, and I wasn't surprised that it was frogs with extra legs. I mean, that's not surprising at all. So I talked with Sue Bryant and Dave Gardiner, and I think it was Dave who actually came up with the idea of putting beads into the developing limbs."

Since Sessions didn't have any cercariae, he couldn't watch them infect frogs and then see what happened. But thanks to Gardiner, he was able to test a slightly different hypothesis: Could a blockage in the developing limb by an object the same size and shape as a metacercarial cyst cause abnormal development? When Sessions implanted beads in developing limbs of both salamanders and frogs, the answer was yes—although the extra toe and foot structures he obtained weren't nearly as dramatic as the tangled masses of extra legs in the wild-caught frogs from Aptos.

I told Sessions I found it puzzling that such seemingly random attacks from parasites could result in something as nonrandom and "orderly" as an extra limb, especially one that appeared almost identical to the normal, primary limb. I didn't think any two frogs would ever be infected in quite the same way, and yet many of the deformities looked very similar.

Sessions was amused by this. It was precisely this orderly response to a disorderly attack that made the case for parasites so compelling, he said. The abnormalities that arise from infection are actually the organism's best guess at what *normal* should look like. Intercalation is a repair program, but it's inherently faulty if the tissues under repair are have been rearranged from their correct positions.

"In some specimens you can't find much order at all," Sessions said, "though in many you do. But this is the way intercalation works. Remember, intercalation is a mechanism to reestablish pattern continuity at the cellular level. So a mirror-image duplication is exactly one of the things you might expect from a random moving around of cells. You randomly move the cells around, but the response is decidedly not random."

When cells in the developing limb got mixed around, Sessions said, they set about remaking a proper pattern. Molecules on the cell surfaces that detected adjacent cells that were out of place recognized that something was missing. "So these cells begin then, in a very orderly fashion, to mitose and proliferate. Their daughter cells take up positional values that are intermediate. This creates new positional confrontations that have to be resolved. Eventually the whole structure works itself out and in the end it becomes nonrandom."

I was following this, barely.

"So basically, intercalation keeps filling in the blanks until order is reestablished," I said. "But since some cells are out of place that can mean an extra leg. Or ten."

"Yeah," Sessions said. "The cells don't care. All they care about is having normal neighbors. They're smart."

Later I spoke with Bryant about the results presented in "Sessions and Ruth," which I knew she didn't think much of. "It's funny," she said, "all the attention that paper ended up getting. It's one of the very few papers that ever came out of my lab that I didn't put my name on."

Most scientific papers have multiple authors, with the principal investigator listed first. Supervising scientists are often among the names that follow and it would be routine for Bryant to have added her name as a coauthor to any of her postdocs' papers. Bryant told me she occasionally withheld her name from papers published by fellows in her lab for one of two reasons. In some cases, she said, she felt the principal investigator needed the credit independently. In others, she believed the data or the conclusions reached were unconvincing. With "Sessions and Ruth" it was the latter.

"I just thought the evidence was weak," Bryant said. "And I didn't like the interpretation. You could find these parasites, but not in what was supposedly the right place in every case. And the structural duplications associated with the bead insertions seemed pretty minimal.

"I don't know. It just didn't do it for me."

———

Sessions wanted to show me what he was talking about. He told me to grab my boots and we got into his minivan and headed out of Oneonta east on Interstate 88. Sessions said he knew of a number of wetlands along the highway that had plenty of frogs and snails. The day was gray but pleasant, and we were one of the few vehicles on the road. Sessions wore a jaunty long-billed cap and seemed merry at the prospect of bringing something interesting back to the lab for my benefit. He kept right on talking.

Sessions said the problem with retinoic acid as a plausible cause of frog deformities was that all of the evidence that suggested this possibility had been obtained in experiments using extremely high concentrations of retinoic acid or one of its alternative forms. I asked him if he thought such a high dose could occur in a wetland exposed to a compound that mimicked retinoic acid.

"No," he said, "I don't see how. I mean, it would take millions of dollars' worth of retinoic acid to get even a small wetland up to the kind of concentration we've tested in the lab."

This was a claim Sessions made over and over again. A few months after our talk, he was quoted in *USA Today* saying it would take "thirteen million dollars' worth of methoprene to cause amphibian deformities in a single pond." Where he got that—methoprene still hadn't even been shown to cause such deformities in the lab—was hard to imagine. Nor could the possibility of bioaccumulation of methoprene or any other toxicant be ruled out. There were many well-documented cases of environmental contaminants—such as DDT and PCBs—that built up to higher concentrations through the food chain or via prolonged exposures to lower doses. Yet some other researchers did agree that there was a considerable difference between the experimental evidence showing that retinoic acid could produce abnormal limbs and any scenario that might occur in the wild. The experimental doses that produced extra or deformed limbs in mice and chicks did in fact involve high concentrations that were administered only briefly during critical developmental stages. When I asked Joe Tietge about this he said the same doses would be lethal to an animal exposed to them "chronically" from the embryonic stage onward. To get the abnormal limb result, the exposure had to be turned on and off at just the right time. This "pulsed" dosing was different from what most researchers envisioned happening in a contaminated wetland, where the exposure would presumably be continuous. In plain language, the question was how a tadpole could live to become a deformed frog if it were swimming around in a fatal dose of retinoic acid, or a mimic of it. The effects of lower concentrations were unknown. "There is no evidence in the literature that tells us a sublethal dose of retinoic acid can produce limb abnormalities," Tietge told me. But the EPA was testing the possibility.

Much of this speculation on effective doses and modes of exposure was only that. One of the intriguing findings of the endocrine-disruption research done in recent years has been the observation that hormone mimics may act at very low doses and in some cases produce adverse effects only at low concentrations.

I put the question to Dave Gardiner. He didn't see any reason at all why a retinoid couldn't be involved in the deformities and, like Ken Muneoka, he in fact believed that retinoids should be considered the prime suspects. In Gardiner's view, the fact that the experimental evidence showing limb deformities being caused by high doses of retinoids in no way argued that low doses might not also do the same. Nor was he bothered by the fact that retinoids had always been tested in pulsed doses rather than over longer periods of chronic exposure. Gardiner answered my question with one of his own, which called for a conclusion I hadn't thought of yet. I learned this was a habit with him, as was his penchant for making every question sound instead like a statement by beginning his answer with the word *so*.

"So the data show that you get abnormal limb development with a pulsed dose, right?" Gardiner said.

"Right," I said.

"So what inference do you make when you see these wild frogs with deformed legs?" he asked.

"I'm not sure."

"Really?"

I thought for a minute.

"Well, I guess the inference is that if it's a retinoid then they are somehow receiving a pulsed dose," I said.

"Exactly," Gardiner said.

Gardiner said he thought contaminant exposures in a natural wetland could be highly variable, especially if the culprit were a pesticide. Pesticides are often most heavily applied in discrete time periods early in the season—just when frogs would be reproducing and developing. Why not a pulsed exposure?

Unbeknownst to the North American researchers at the time, there *was* evidence that suggested chronic exposures to sublethal doses of retinoids could affect limb development. And it was Gardiner who eventually found it some months later. In the early 1970s, a team of Italian scientists had conducted a sequence of experiments in which they tested the effects of vitamin A—the parent compound of retinoic acid—on amphibians. The initial studies demonstrated that vitamin A was a potent teratogen that caused profound disruptions in early development, particularly in the nervous system. These malformations were always fatal, and the researchers—Giovanna Bruschelli and Gabriella Rosi—wondered what would happen if they dropped the dose.

Their results, which were published by Perugia University in 1971, showed that progressively reduced concentrations of vitamin A adiminstered to *Bufo vulgaris*, the common toad, allowed the animals to survive for longer periods. As the dose dropped, organ systems that developed later were affected by the treatment. Bruschelli and Rosi described one of these deformities, which was evidently unprecedented, as "a particular anomaly consisting in the bilateral and symmetrical duplication of the hind limbs."

Sessions and I trundled back up the steps to his lab. He was carrying a bucket we'd filled with muck and snails at a small pond just below the road shoulder of I-88 a few miles out of town. Sessions had needed only a few scoops with his net to get a goodly number of snails, which he sorted out of the weeds and mud with heavy, long-handled forceps. Back on the road he continued on a few miles before finding an "Official Vehicles Only" turnaround in the median and swinging through it nonchalantly to head us back into Oneonta.

"Let's have a look at how these guys operate," Sessions said. We were now at the microscope table again. Sessions had switched on the microscope that was hooked up to the video monitor. He took a tadpole, one of the wood frogs, from his tanks, and dropped it into a beaker of anesthetic for a couple of minutes. Just long enough to immobilize the animal. This made it easier to watch under the microscope when the parasites hit them, he explained. "Otherwise they writhe around in agony and you can't focus on them," Sessions said.

I looked at him.

"Well, I'm kidding, sort of," he said. "Actually, I would think it is rather painful."

While the tadpole quieted, Sessions retrieved a couple of snails from the bucket and crushed them open with forceps in a small dish of water. This, he explained, was a little gruesome, but it was the only way he knew of to extract the cercariae—which were now invisibly swimming around in the dish, along with small floating blobs of gray and pink snail tissue. Sessions transferred the tadpole to a dish small enough to go under the microscope, and gently moved it around until we could clearly see an enlarged view of the developing hind limb region on the video screen. The legs were still incomplete and appeared as thin, lightly pigmented stumps lying close to the flank near the base of the tail.

Sessions drew up some of the liquid from the snail dish with a pipette and dropped it into the dish with the tadpole right next to where we were looking. Watching the video screen I saw some tiny, almost transparent obloids moving around in a jittery, random way in the brightly lit water. At first they reminded me of the spots that sometimes drift across your field of vision on a sunny day. But when I looked closer I could see that each had a complex internal structure and a short, vibrating tail that propelled it. These were trematode cercariae. Sessions swirled the water around a bit to stir some of them closer to the tadpole. This, he conceded, was help that they wouldn't get in the wild. But in the interest of time it would help me to see what happened when they encountered the tadpole.

For the next hour or so, that is what we watched. Over and over again, the same meeting between parasite and tadpole took place: The cercariae would bump along the skin of the tadpole, nosing their way, tails pumping furiously. Some did this for a while before giving up and swimming away aimlessly into the void. But others kept at it, worrying the tadpole until finding some imperceptible portal and squirming their way in beneath the skin. As the outer layer of skin closed behind the burrowing parasite, the cercarial tail dangled outside, continuing to kick until it abruptly broke free and fluttered off on its own. The cercariae, meanwhile, took up immediate residence just under the skin, curling into tight little balls. They didn't seem to show any particular preference for where they lodged.

I asked Sessions how he'd put the beads into his experimental animals and he showed me some extremely fine forceps that were so sharp the tips literally vanished before my eyes when I tried to focus on them. "We just jammed in as many beads as we could fit into the limb buds," he said. This was not a careful surgical procedure, Sessions said. Quite the opposite. The idea had been to mimic the kind of gross tissue destruction Sessions imagined occurred when the parasites roughly bored their way into the tadpoles.

It seemed like a good time to pose the main question I'd come to Oneonta to ask Sessions: Why the double retreat—first at Duluth, where he'd hedged on parasites as the sole cause of the deformities, followed by the subsequent backflip at Shenandoah where he insisted it was a waste of time to study any cause other than parasites? Sessions grimaced. He said he wished he hadn't equivocated at Duluth, but he'd been thrown off by what he heard there. It wasn't until later that he realized he'd seen a lot more than frogs with multiple legs at Aptos, that in fact he'd seen a bit of everything there.

"Actually, when I first heard about the Minnesota situation I immediately suspected a chemical substance," Sessions said. "That's the first thing everybody thinks of. You see a screwed-up animal in the field and that's the conclusion to jump to."

Even so, Sessions told me he'd arrived at the Duluth meeting feeling pretty confident he'd convince everyone to set aside their initial judgments and see that parasites could be the cause of the deformities. But he also said he'd tried to go with an open mind and suddenly found himself uncertain that parasites would account for everything that was being observed.

"You know, the only person who really showed frogs that looked like what I'd seen was Dave Hoppe," Sessions said. "He showed pictures of frogs with multiple legs. And I thought those were being caused by parasites. And I am still certain they are."

What threw Sessions off, he said, was the prevalence in the reports of so-called cutaneous fusions, as well as the many missing legs. Sessions had since grown dismissive of the missing-leg phenotypes, which he now referred to as "stubby amputees," a description that presumed a cause for which there still wasn't any hard evidence. If Sessions thought that all the missing limbs were in fact the result of trauma—predator attacks, lawn mower accidents, whatever—then he was discounting the great majority of deformities reports out of hand.

"It is really hard to call animals with missing legs deformed frogs," Sessions told me. "They're just frogs that are missing pieces of their limbs." We both laughed at that for a moment. Then Sessions was serious again. "It's something I didn't dare say at the Shenandoah conference, but that's exactly how I feel. They are not deformed frogs."

Sessions said he hadn't remembered seeing many of the morphologies that were being found in Minnesota and Canada among his earlier specimens, but that when he went back to his original data from Aptos, there they were, all present and accounted for. Skin fusions. Twisted limbs. Missing limbs. The whole ugly repertoire of abnormalities. "When I looked at my frogs again I began to fill in the holes and narrow the gap between my findings and what was now being seen," he said. "I did have a lot of those things listed in my tables of deformities. I just didn't remember. But after I looked through the frogs again I saw that I could explain a lot more with parasites than I thought I could."

The big problem, however, was the frequency of missing limbs among the Aptos specimens. Sessions did see this morphology—but not often. Of 280 Pacific tree frogs he examined, only three had missing hind limbs, just a shade over 1 percent. That would be about equal to what had by now become the accepted background rate. Sessions told me the high, double-digit rates of missing limbs reported in other places were indeed puzzling—whether you thought they were deformities or not. And he didn't think parasites were the probable cause. "The one thing I couldn't explain were all those missing limbs," he said.

But of course that was exactly the problem to explain. Missing legs were the main data—extra legs were just the headlines. Sessions conceded that the frogs from CWB and from several of Martin Ouellet's sites that had extra legs might just be "outliers," relatively rare episodes of parasitism that didn't have anything to do with the apparent outbreaks of missing limbs. Sessions told me that his inspections of some missing legs strongly suggested trauma, probably from predators, because he sometimes saw a partially successful regeneration attempt at the tip of such a limb. But not always. And he'd said he simply couldn't say what he thought of frogs from Minnesota that had missing limbs or anything else because he never got to look at them. Sessions said the Minnesota researchers had repeatedly refused to send him any frogs.

"I wouldn't care," he said, "it wouldn't bother me if they would just clear and stain them and then show the results. But it starts to make me suspicious that they haven't done that. I told them it was important. So did Ken Muneoka."

I told Sessions I wondered why this was so. I suspected that his belligerence at Shenandoah was a big part of the problem, although I didn't say this to him. Sessions said only that he was puzzled, too, but thought it was possible that the whole frog investigation was being manipulated—and important evidence ignored—so as to promote further research funding. In essence, he seemed to think other investigators were tilting their hypotheses toward a chemical contaminant in an effort to catch the endocrine disruption wave.

This in itself wouldn't have been terribly surprising. Fads and trends exist in science, as elsewhere. Researchers naturally look for ways to make their investigations current and in step with funding priorities. Sessions's reservation was essentially the same one that Mike Lannoo had offerred to me a few months earlier. The difference, though, was that Sessions believed people also wanted to actively suppress the parasite hypothesis while promoting the idea that chemicals were to blame. He said at least one "respected colleague" had urged him to keep the parasite business "under wraps" for fear of jeopardizing the grants that would more readily be available to study chemical contamination.

"If you just want to study the natural history of trematodes and garter snakes and frogs," Sessions said, "then you're going to have a hard time getting grant money. A really hard time. If, however, you're talking about a pesticide that's causing developmental defects in frogs using a basic biochemical pathway that also exists in human embryos—then you're in the money."

Nobody, Sessions added, had yet shown even a sliver of evidence that would make chemicals a viable suspect in the deformities, which in his view left parasites as the "dispassionate" explanation for the evidence. This was a typically bullying statement from Sessions, lumping all alternative points of view together as hysterical. But I could see that's how it looked to him and that's what he believed, absolutely. "I can't really help it," he said.

Martin Ouellet sat on a heavy log looking at frogs and making notes as he picked them up, one by one, from a bucket at his feet. It was another lovely September day in the lingering summer of 1997. We were in the shade at the head of a hundred-yard mudflat by the mouth of a small stream on the southwest shoreline of Lake Hertel, which lies serenely about a third of the way up Mount St. Hilaire in Québec province. The mountain itself rises from the St. Lawrence River plain about an hour east of Montreal. Mount St. Hilaire is one of several domed, extinct volcanoes in the area—Montreal itself is another—and stands as an ecological island in the midst of the sprawling farm fields and suburban tract housing developments on the valley floor below. McGill University owns a large, undeveloped reserve here and Ouellet had adopted it as an important study site.

His main purpose on this visit was to survey the area for *Rana palustris*, the pickerel frog, a near cousin of the northern leopard frog and the rarest frog in Québec. To Ouellet's delight, we'd caught a couple dozen juvenile pickerel frogs basking on the mudflat. Ouellet worked methodically and at a brisk pace. Since he expected to make subsequent collections here, he marked each of the frogs caught this day by clipping off the tip of the third toe on the left foot so that he would recognize these specimens among any he picked up later on. Toe clipping is standard field practice for herpetologists monitoring amphibian populations, at least for animals that still possessed toes to clip. I'd seen it done before, though never so efficiently and caringly as Ouellet performed the job. Most herpetologists simply snip off part of a toe with nail clippers dipped in alcohol and chuck the frog back in the swamp. Ouellet used small, pointed surgical scissors in a two-part operation that took him just seconds to complete. First he cut off the tip of the toe. Then he drew back the skin from this cut with the fingers of the hand in which he held the frog, so that the end of the bone—a fine, white filament—was pushed outward and exposed. Ouellet snipped this tiny segment of bone off flush and pulled the skin back over it, pinching it together and sealing the wound. He then swished the frog in the water of the creek where it ran under the log. He told me this technique reduced the chances of infection and gave the frog some protection over the amputation so it wouldn't chafe and become inflamed as the frog moved around.

It was relaxing to watch Ouellet work—he had the herpetologist's Zen in the way he handled the frogs. And these frogs were in good shape, as Ouellet expected they would be at this far remove from agricultural contamination. But they were not all perfect.

"Ah," he said, holding out a frog at arm's length. He'd scarcely more than glanced at it. Ouellet passed the frog over to me. It was much like a leopard frog in appearance, only somewhat smaller and with spots that were more rectangular and orderly in their arrangement. The frog's underside, near the groin, had a bright yellow tint.

"And what do you see?" Ouellet asked me. "Is it normal?"

I inspected the frog carefully, extending the legs to look for asymmetries or skin webbings. I counted the toes. Everything looked good. I held the animal up and looked at it head-on, studying its ironic frog smile, the 350 million-year-old inscrutable smirk, which always seemed such a cruel joke in specimens that were deformed.

"I believe it is normal," I said, handing the frog back to Ouellet. I was all but certain this was the wrong answer, but I really couldn't see anything amiss.

"Well, no." he said. "You see how easily we can overlook something? Here, have a look at this toe." Ouellet held out the right rear leg and prodded at the second digit—which was in fact missing the tip beyond the last joint. Unless you've looked very closely at a frog's toe you can't imagine what a small wisp of tissue was gone. Ouellet had seen it right way. Ouellet's powers of observation were legendary among his coworkers. So was his tenacity in gathering specimens. On more than one occasion Ouellet had chased and caught a snake that was in the process of swallowing a frog and forced it to regurgitate the animal so he could count it in a sample. I looked blankly at the shortened toe.

"A deformity?" I asked.

"Ah, no again," said Ouellet. "I'm thinking this is from a fish." He nodded at the lake as he put the frog into the release bucket, then made kind of a nipping motion with his hand. "This lake is having plenty of fish. This toe, I think it is a trauma, you know?"

In fact, I admitted I was a little unclear on how extensive Ouellet believed traumatic amputations were in wild frog populations. At the

Duluth meeting he had seemed to indicate that he presumed many short-ened or missing limbs were the result of trauma. He'd mentioned preda-tors. He'd mentioned farm equipment. Now he told a story about finding a large number of amputated limbs on frogs in a wetland adjacent to a field that was being cut for hay. He'd said one had to be skeptical about missing limbs being developmental problems.

Ouellet told me that I'd misunderstood slightly what he'd been say-ing in Duluth. He thought traumatic amputations occurred, but rarely at high frequencies. Moreover, such amputations were usually pretty obvi-ous. So you had to rule out trauma in the case of missing limbs in order to call them true deformities. But often you could do just that. There were usually signs of bleeding or infection or scarring that gave away trauma as the cause of a missing or shortened limb. But what Ouellet had seen throughout Québec was different. This mass ectromelia was *not* a case of mass trauma, he told me emphatically. What he saw in agricultural ponds that were exposed to pesticides was a developmental disorder. It was clear-cut. Much of the ectromelia he observed was not simply missing or partial limbs, but limbs that were strangely shrunken or "atrophied" in appear-ance. In the years since his first studies he had compiled much more extensive evidence to support his claims. This summer alone, Ouellet said, he added another 10,000 to 15,000 frogs to his total sample. This dwarfed the numbers being compiled in Minnesota and elsewhere in the United States. Ouellet's data now included intensive sampling at more than a hundred sites. His minimum sample size in every survey was a hundred frogs, but the target was three hundred and at some places he had many more than that.

"What was the overall percentage of deformities that you found?" I asked.

Ouellet said the numbers were still just a gut feeling, because he hadn't yet computerized the data, but that on average frogs at his control sites exhibited between zero and 1 percent abnormalities—true deformi-ties, he said, not amputations. Most of the time all of the frogs at the con-trol sites were normal.

"In the pesticide sites I sometimes find zero percent abnormalities again," he said, "but in others it's as high as seventy-eight percent." Ouel-let said he had at least five sites where the deformities rate topped 70 per-cent and that the average frequency in pesticide-exposed wetlands seemed to run around 10 percent. He added that he also saw a much wider range of deformities in the agricultural ponds.

"So you see more deformities and more severe deformities in pesticide sites," I said.

"Oh yeah," he answered quickly. "It's clear-cut. Finding one deformed animal in a control site—you know, it's not so informative. But finding so many in a pesticide site is informative. This is why it's important to have a large sample size."

Ouellet said he had greater numbers of mink frogs and green frogs in his surveys, but only because those were the most common frogs he encountered. In a few places, he saw the confused mix of deformities Dave Hoppe was reporting from CWB, but mostly what he saw was missing limbs.

"I have a lot of other stuff," he said. "Like cutaneous fusions. A lot of muscular rigidity. All kinds of reductions of the bones and the limbs."

Ouellet said he estimated the rate of traumatic amputation he saw in the field ranged between 1 percent and 5 percent of the frogs he caught, and that this didn't vary noticeably between control sites and agricultural sites. Once in a while he ran across high rates of amputation, usually of the digits, but that visible bleeding or swelling made it it clear these were "fresh" wounds.

I thought this was a startling finding—although Ouellet seemed quite casual about it. None of the other field biologists I'd spoken with routinely came across missing limbs, traumatized or otherwise, and most of them—Andy Blaustein was a notable exception—had also been skeptical that predation could account for such abnormalities. "If your sample size is large enough," Ouellet reminded me, "you will find everything. After six years, I'm still finding new kinds of deformities." Apparently he also continued to find mounting evidence that predators in fact do sometimes take off only legs or parts of legs. I asked him if he found fish in his control sites. He said he did. Same for the pesticide sites.

"It's about the same everywhere," he said. "My correlations are not done, but I think there is an association with the presence of fish or other kinds of aquatic predators."

So the chances that a frog with a missing limb was the result of predation would be about the same whether it was in a farm pond or one undisturbed by pesticides?

"I would say yes," Ouellet said.

Like Dave Hoppe, Ouellet had come across skin webbings most often in mink frogs, but as he increased his sample sizes they turned up in other species as well, sometimes in the same places where there was an abundance of missing legs. Ouellet said he was certain that a cutaneous fusion was a developmental problem, not the result of any conceivable trauma. So were many of the other new deformities he was coming across. This summer, Ouellet said, his fieldwork had turned up any number of strange abnormalities, which even he hadn't seen previously. "There are frogs that are monsters," he said. "They are impossible to describe. Only a picture can do it because it's a mess. A total mess. And I'm having a lot of total mess. I have, sometimes, a piece of skin, skin like a string, that starts from one leg and goes to the shoulder. Things like that. A piece of skin. You can't even relate that to anything, you know? It's weird. I found all kinds of stuff. I found deformed mouths a lot this year."

We were riding in the car when Ouellet told me this. He rummaged through his backpack, which was always at hand, and asked if I was hungry. When I said I was he produced a foil envelope containing two strawberry Pop-Tarts, ripped it open, and passed one my way. We sailed along, chewing thoughtfully.

"Could it be parasites?" I asked.

"Well, I know one guy who will say yes," he said.

Ouellet grinned. "Me, I will say no. But I don't know. I am curious to see if the bones beneath the skin are affected, so I will check for parasites. In all my truly deformed frogs I want to rule out parasites. So I'm checking on that. I think parasites are great. But they don't explain what I'm seeing. There are snails everywhere. Sometimes ten times as many at sites where there are no pesticides. But I do not have mirror-image duplication. I do not have a lot of extra legs. Parasites do not explain all my reduced limbs, all my missing limbs, all my cutaneous fusions. I don't see a chance that these are caused by parasites. Parasites seem to be involved in extra legs, from what we now know. A lot of mirror-image duplication. I think out of more than twenty thousand frogs I have recorded two mirror-image duplications. So parasites might explain two of my frogs. For me, I'm quite happy Stan is involved because he forces us to think about how natural events can be a factor. I think that is important."

———————————

I spent three days with Ouellet, at his office at McGill and crisscrossing Québec province to visit a number of the sites where he was monitoring frogs. On a swing through an agricultural area west of Québec City, we picked up Ouellet's Canadian Wildlife Service (CWS) supervisor, Jean Rodrigue, and together we inspected wetlands on several farms. It was beautiful country, with hard-packed soil that was densely planted with a dizzying array of crops: feed corn, sweet corn, strawberries, raspberries, asparagus, potatoes, wheat, alfalfa. The potato crop was being readied for harvest and had been treated with herbicides to kill the aboveground parts of the plants, which now looked like rows of little brown Christmas trees wherever we saw them.

The wetlands we went to varied by type. Some were ordinary reedy ponds or low, swampy areas in the middle of cultivated fields. Others were manmade, steep-sided rectangles that looked deep and clear, sometimes with a blue or green tint to the water. None seemed to have much in the way of emergent vegetation. These had been constructed out of the way along the edges of working fields. They ranged in size from perhaps fifty feet by twenty to roughly twice that. Rodrigue explained that in the 1970s and 1980s the Canadian government had reimbursed farmers for the construction of these ponds as replacements for natural wetlands that were plowed under. They were used primarily as water sources for fire control and irrigation. The presence of wildlife in them was entirely accidental. Farmers occasionally tossed fish into the ponds, and in some the fish flourished.

The contamination of these wetlands by agricultural pesticides was quite inevitable. Rodrigue said the ponds were polluted by all the usual inputs from the surrounding countryside: overspraying, runoff, drain tile systems, and even tainted groundwater seeping in from below. Many farmers dumped empty pesticide containers into their ponds as well. Previous studies had shown that pesticide levels in such wetlands often exceeded the amounts recommended to protect aquatic species. All of which, said Rodrigue, frustrated the government's efforts to reduce pesticide residues on produce. Canadian law specifically bans the application of pesticides to food crops after certain dates so that freshly treated plants will not be harvested and sold. But nothing in the law speaks to when pesticide-contaminated water can be pumped out of ponds and back onto the fields—something that Ouellet and Rodrigue said happened all the time.

I wondered about Rodrigue's role and the power of the CWS in Ouellet's work. It was clear that Ouellet had access to a great many wetlands on private property, and that as a result his data sets were fundamentally different from what was being collected in the United States—where study sites were limited to public lands and a handful of private properties whose owners were curious enough about the deformities to invite the researchers in. Dave Hoppe had often bemoaned the fact that nobody in authority in Minnesota ever "flashed a badge" that would admit researchers onto private property so they could study farm ponds of their own choosing. Even the dairy operation right next door to the Bocks' house had gone uninspected.

Ouellet told me the CWS had provided him a polite but insistent foot in the door with Québec farmers. Rodrigue never forced farmers to let them on their property, but the CWS had an aggressive public relations department that prearranged for Ouellet's visits and helped to preserve the study sites he set up on private land. It helped, Ouellet confessed, that few farmers in rural Québec speak English or read English-language newspapers. The deformities problem, he said, was being almost completely ignored in the French-language media and therefore nobody was concerned that their farming practices might be called into question.

The differences between Ouellet's studies and those ongoing in the States were profound, and in a way his results seemed to argue against some of the assumptions that had been made in Minnesota and elsewhere. Investigators in Minnesota were actively pursuing the possibility that farm pesticides had become ubiquitous contaminants, with potential long-distance effects even in wetlands that did not appear to be directly affected by agriculture. But Ouellet's data suggested the connection between direct exposure to pesticides and the presence of deformities was strong. His evidence that deformities were generally absent from nonagricultural sites was equally compelling. Ouellet found deformities on farms. He didn't find them elsewhere.

Ouellet also agreed with Stan Sessions on a key point. He believed a site like the Bocks' lake had to be regarded as atypical. He'd found Dave Hoppe's data from CWB just about the only useful field results from Minnesota, but he said focusing on one site with an unusually dramatic deformities outbreak was risky. Ouellet thought his approach—a comprehensive, random survey across multiple land uses—was an essential first step in studying amphibian deformities that had been skipped in the U.S. investigation.

This was undeniably true, and it was surprising that the Americans continued their efforts in the same vein after the Duluth meeting—as if Ouellet's findings did not exist. This may have only been partly because it was easy to ignore work done in another country. The other problem was that Ouellet was very slow to publish his findings and had gotten nothing out since his original two-year survey in the early 1990s. Still, Ouellet told me he was disappointed in the lack of reliable field data from the United

States. Based on what had been reported, Ouellet said, U.S. investigators could not even make the case for an increase in deformities, let alone say anything definitive about whether chemical contamination could be implicated in the outbreaks. "I don't even know if you have a deformities problem there," he said impatiently. "In the United States it's always the same couple of ponds that you study. That's it. No big survey."

Ouellet was only emphasizing a point. He made it clear to me on many occasions that he very much believed there was a "problem" in Minnesota, probably the same one that was occurring in Québec. But until the Americans stopped looking only at the easy-to-get-to sites on public lands and the handful of private locations where they'd been invited in for a look, the field data was of little value. "If you are not finding deformities in agricultural sites subject to pesticides," he said to me once, "then perhaps someone will bring me there and I will find them for you."

Ouellet had undertaken an ambitious study in 1997 to test the effects of pesticides on frog larvae. At a field station McGill operated in suburban Montreal, Ouellet had set up twenty-four barrel-sized tanks and begun rearing trials on wood frog tadpoles. There were 400 to a barrel, 9,600 larvae in all. Ouellet tested various concentrations of two pesticides: the insecticide endosulfan and atrazine, the most widely used herbicide in North America. The doses ranged from 10 percent to 20 percent of the concentrations typically measured in agricultural ponds. Ouellet hoped this low-level chronic exposure would reveal any sublethal teratogenic properties of the pesticides.

But the experiment was inconclusive. When I visited the setup in September, only about 70 of the nearly 10,000 tadpoles had metamorphosed. Some had died; most were still swimming around in their tanks, apparently healthy enough but also failing to progress out of their juvenile stage. Ouellet thought the pesticides could be a factor in this developmental retardation—wood frogs typically metamorphose in early summer—but he was also looking at his feeding regimen as a possible cause. Other than the slowed development, tadpoles exposed to atrazine didn't seem to show any negative responses, while the ones treated with endosulfan sometimes died at early stages and in some cases developed a temporary paralysis from which they eventually recovered.

Ouellet was frustrated with this indeterminate outcome. His research had only a fraction of the financial backing of the U.S. effort—something on the order of $20,000 Canadian. He couldn't afford to waste time or resources. And he could not understand the Americans' refusal to see the need for a wider field program. The U.S. investigation seemed to be getting funded for being incompetent.

He put it like this: "If you are going to scream, you need to have the data to justify it."

Bad Boy

MARTIN OUELLET'S HARSH ASSESSMENT OF THE U.S. DEFORMITIES investigation soon enough paled next to what was being said about it by the actual participants. The sharply negative response to the MPCA's announcement of the FETAX findings turned out to be only the first volley in an escalating war of words over results that were rushed into public release without the usual arduous checking and double-checking that is supposed to be integral to the scientific method. Enemy camps formed. The investigators savaged one another to anyone who would listen. Judy Helgen, whose own uncooperativeness had already been felt by many, walled off the MPCA's efforts. Anyone outside her shrinking inner circle was deemed "not a team player." The agency's formerly long list of collaborators dwindled.

A few investigators stood out of the line of fire, appalled at what was happening. Mike Lannoo, who'd been recruited by the MPCA to perform high-resolution X-rays on the deformities, was distressed at the stinging rancor that suddenly enveloped the inquiry and the loss of objectivity that came with it. In early December of 1997, just over a year after the Duluth meeting had come to a close on a friendly, collegial note, Lannoo found

himself speaking before many of the principals at a meeting in North Carolina and pleading with them to set aside private agendas and to suspend
further personal attacks on one another. "Let's remember," Lannoo said
ruefully, "that we are scientists."

Lannoo's plaintive call for a return to civility and the rigors of peer-
reviewed research was a nice try, but it did little to calm the fevered
exchange of claim and counterclaim. Science may be objective, but scientists are as prone to human weakness as anyone. As rational imperatives
and irrational impulses mixed freely in the deformities investigation, so
too did the sins of envy and hubris.

––––––––––

The FETAX results were headed for a rude reception long before Judy Helgen decided to go public with them. Earlier that summer, on one of several
visits I made to the EPA in Duluth, Joe Tietge had hinted to me that some
people working on the deformities problem might be on the wrong track
and appeared to be moving toward hasty conclusions. He didn't mention
FETAX specifically, but he did say there had been a sudden chill in relations between the EPA and the NIEHS-MPCA partnership. Tietge was, as
usual, judicious in choice of words.

"I do have a reason to believe that we need a little bit better interagency communication," he said. "I've brought it up. We'll see where it goes."

I asked him specifically if he'd had conversations with the NIEHS
recently.

"No," he said, "and that's one of the reasons I think some sort of
interagency working group would be appropriate." Tietge conceded that
Jim Burkhart had, in fact, phoned him not long ago. He said he'd not had
time yet to return the call. But he didn't sound hopeful about renewed
contact with Burkhart's group. "We tried to collaborate with the NIEHS,"
he said. "It didn't go anywhere."

Tietge seemed to know more than he was willing to say at the time.
And he did. The truth was that cooperative efforts between the EPA and
the NIEHS-MPCA investigation had shut down completely—evidently
over a dispute between Tietge and Judy Helgen over the sharing of water
samples. Tietge had requested duplicate samples from sites the NIEHS was
analyzing. But Helgen refused, and told me later that she did so because
Tietge declined to tell her what he intended to do with them. Tietge said
Helgen demanded what amounted to a written work plan—which he

found unnecessary as this had never previously been requested. He also thought it was insulting.

The reason Tietge wanted the water, which Helgen may or may not have guessed, was that the EPA was running its own FETAX studies in Duluth. The purpose was less to see what might be causing frog deformities than to look for shortcomings in the assay itself. The drawback with FETAX, Tietge told me, wasn't so much that it used *Xenopus laevis* as a substitute for native frogs. Rather, it was that the assay tested for effects only during a brief window at the very beginning of development. The EPA, he said, had considered using FETAX in its work on the deformities problem, but decided against it because the assay did not achieve an endpoint that would even remotely parallel what was being seen in the field.

The assay is standardized and relatively easy to perform. It's a straightforward exposure and response exercise. Freshly laid eggs from the African clawed frog are harvested within a few hours of fertilization and their protective jelly coating is removed. They're then placed in a solution containing the mixture being tested—water or sediment extracts in the case of the Minnesota work. The eggs hatch and become tiny, free-swimming larvae. After 96 hours of exposure the larvae—which are yet only half an inch long or less—are microscopically examined. If a significant number of them have died, this is considered an important indication that a toxicant is present. Surviving larvae are checked for an array of early stage developmental abnormalities that have been extensively catalogued as responses to teratogenic compounds. These abnormalities, which typically include malformations of the skull, eyes, face, mouth, nervous system, or gut are, of course, "deformities." Just as embryonic *Xenopus* larvae are technically "frogs."

But they are hardly deformed frogs in the sense anyone understood the term in the context of what was happening in Minnesota. At 96 hours, *Xenopus* have begun to form limb buds, but they are a long way from having legs. Tietge told me the EPA regarded *Xenopus* as a useful and important research species that figured in thousands of scientific papers published every year. But the endpoint everyone was interested in was abnormal legs, and FETAX didn't show what happened to developing limbs. Extrapolating from a different sort of earlier developmental defect like a twisted gut or a malformed eye was inherently risky, Tietge said. Identifying such malformations could be a subjective proposition. Deciding what frequency of abnormality consituted a "positive" FETAX reading could likewise be a judgment call. Embryonic *Xenopus* were sometimes sensitive to nontoxic factors. Tietge said their development could be affected "if you look at them the wrong way."

"The problem you've got with the standard FETAX protocol is understanding what it means," Tietge said. "You have to be very careful how you interpret it. As far as implementing it as a general developmental screen—I don't think it's accepted for that purpose."

In the case of the Minnesota water samples, Tietge said, FETAX results only seemed relevant if you were willing to risk an "extreme interpretation" of the data. This wasn't the sort of thing, he added, that ought to be the basis of a press conference: "We think hard data should be presented through normal channels like a scientific meeting or a peer-reviewed publication." What Tietge didn't say—what everyone knew but did not acknowledge—was that the announcement to the public that "deformed frogs" had been created in a lab via exposure to Minnesota water was at best a carefully parsed half-truth.

The condemnation of the FETAX announcement was nearly universal in the weeks following the press conference. Mike Lannoo renewed his complaint that *Xenopus* was "phylogenetically remote" from any native frog species. The fact that *Xenopus* is a frog only confused the issue. Andy Blaustein was also furious. Going public with the results, he told me, was "alarmist and premature."

"They're looking at a totally aquatic species from Africa with a very different physiology," Blaustein said, fairly screaming into the phone when I called him. "And they're looking at it after four days of development, before it even has legs!" Stan Sessions was stunned. How could anyone seriously believe that well water in Minnesota was contaminated by something that caused deformed frogs in surface wetlands? The implications of powerful teratogenic compounds polluting groundwater, he said, was an environmental nightmare too horrific to even contemplate.

When I spoke with Hillary Carpenter at the Minnesota Department of Health—which has primary responsibility for well safety in the state—he told me his department had argued vehemently against making the FETAX results public during a briefing from the MPCA just prior to the news conference. But he allowed that Judy Helgen seemed genuinely concerned that the findings were about to leak—another *Nightline* program was imminent, with or without a press conference. Ironically, it was on *Nightline* that a key question about the implications of the FETAX study was raised. During his closing chat with correspondent Chris Bury, host Ted Koppel noted that well water pumped from a deep aquifer would have "come from someplace else" that wasn't necessarily in proximity to the types of deformities being studied. Why should there be any connection?

But Helgen felt certain that providing safe drinking water while the facts got sorted out was the only responsible course of action—and also that the best chance they had to manage public reaction to the findings was to put out the information on their own terms. Arguably, it worked. News of the FETAX results and the bottled water distribution was widely reported in the state and the investigation seemed to move on without a hitch.

In late October, however, the EPA turned up the heat. And it did so by *confirming* the FETAX results.

———————

Joe Tietge couldn't get water samples from every place the MPCA and the NIEHS were working, but he had no trouble collecting surface samples from CWB, and these were subjected to the FETAX procedure at the Duluth lab. Tietge added an intermediate step to evaluate the samples for general toxicity by running the assay in a large volume of water. Normally, if a toxicant is present in a sample, increasing the volume while keeping the number of organisms exposed in the assay constant will produce a greater effect. This didn't happen. Meanwhile, at the standard volume Tietge told me they got exactly the same results that the Stover Associates in Oklahoma had obtained and forwarded to Jim Burkhart and Judy Helgen. But the EPA did not believe the craniofacial and gut abnormalities they observed in the assay were caused by the presence of anything teratogenic in the water. Rather, they determined that the malformations were the result of something that was missing from the water, which was unusually soft. When Tietge and his colleagues added calcium, magnesium, sodium, and potassium salts to the assay in amounts that are recommended for the rearing of *Xenopus*, embryos grew normally in the corrected water. The natural "ion imbalance" of the water at CWB was a commonly encountered characteristic of Minnesota water and was also a factor that was known to interfere with bioassays. Plus, it was completely benign from a human health standpoint. On October 28, the acting director of the EPA's Duluth lab sent a pointed letter to the Minnesota Department of Health explaining the ion imbalance issue and offering the EPA's assistance to the state in performing the "relatively simple biological and analytical work" needed to assess the situation. The implication that the MPCA and NIEHS had bungled the FETAX study couldn't have been clearer.

"Results don't mean anything if they aren't interpreted properly," Tietge told me. "Anybody with a tropical fish aquarium knows that if you fill it with tap water it will kill the fish. That doesn't mean your tap water isn't safe to drink."

Down in North Carolina, at the EPA's National Effects Lab, Tietge's boss, Gil Veith took an even dimmer view of what had happened. Veith thought the NIEHS had hastily served up the FETAX results as a "deliverable" to Congress as part of that agency's continuing effort to increase its funding at the EPA's expense. When I spoke with him, Veith was blunt about his disappointment in the lack of cooperation between the two agencies and what he saw as the hurried and innaccurate findings that resulted.

"We had agreed to work together," Veith said angrily. "But in fact there hasn't been a collaboration. We offered to share water samples. They declined. They briefed us on the FETAX results after the press conference, but they never asked for our input prior to the announcement. Most agencies would have sought a consultation with us."

Veith could not contain himself.

"The NIEHS acted irresponsibly in a rush for headlines," he said, "and in the process they overlooked some very basic rules for running bioassays. Now federal scientists are going to look like idiots. Even those of us who are right. And it's not going to help the image of science."

Veith had no doubt that it was the EPA that knew what it was doing. The NIEHS, he said, didn't have much experience with aquatic bioassays and it was evident that they had botched this one.

"There is going to be a right answer to this," Veith told me. "I think the results from Duluth are as clear as you can get."

Joe Tietge seemed to find the whole episode painful and hard to talk about without saying things that were damning of other people's research efforts. This was not his style and it showed. The MPCA, he told me sadly, "had no solid reason to go public" with the FETAX findings. Even though the EPA had looked at only surface water from CWB, he didn't see anything alarming about the results Burkhart's group had obtained elsewhere.

"You could probably take tap water from almost any county in Minnesota and get results like this," he said. In Tietge's view, the MPCA and the NIEHS had a more fundamental problem than misreading a bioassay. Their assumption of a linkage between deformities in the field and totally different developmental problems in the lab was simply too great a leap. "In science, spurious correlations happen all the time," Tietge said. The association of one thing with another often turns out to be meaningless and, he added, "is one of the weakest forms of evidence to support a hypothesis."

Burkhart and Helgen were understandably distressed by the EPA's attack on their work. When I called Burkhart to get his reaction he sounded weary but somehow managed to remain good-humored about it. He didn't agree that things were nearly so black and white. Since the original announcement, the NIEHS and the Stover group had continued to run the assay on more samples and were laboring to analyze results from nearly fifty different sites now. Burkhart admitted that the new data didn't seem to support the initial "concordance" between postive FETAX results and locations with deformities. But he said he was puzzled and disappointed that the EPA was condemning their efforts when Duluth had only performed the assay on surface water from a single location.

"We're still seeking the truth," Burkhart told me. "I'd be loath to make any kind of generalization based on one sample." In any case, said Burkhart, the NIEHS had never claimed to have found a toxicant in the water. The FETAX studies were only one component of a much larger research effort.

But Burkhart was less willing to defend the decision to make the results public. He said he was unhappy that his agency had been forced to announce the FETAX findings before they had fully interpreted the data. "We had no intention of going public until we were further along," Burkhart told me. "But the MPCA insisted and we had to respect their call even though we didn't have all the answers."

George Lucier said much the same thing to me. The NIEHS deferred to the MPCA's decision that this was a potentially important public health issue that demanded a full and immediate disclosure of the findings, even though they were preliminary. Lucier also disputed Gil Veith's complaint that the NIEHS was uncooperative and raised an objection to the EPA's ion imbalance analysis. The Bocks' well water, Lucier noted, had normal ion levels. Yet it too produced abnormal development in FETAX testing.

If Judy Helgen had had any second thoughts about going public, she didn't betray them when I talked with her. In fact, she told me that she felt "very good" about releasing the information. She also confirmed that the decision to announce the results had been made solely by the MPCA. She said she'd make exactly the same call again under the same circumstances. "As scientists," Helgen said, "none of us wanted to go public. We're already doing too much of this work in the public eye." But she didn't think she'd really had the option to treat the findings as a secret, and believed that "as a public agency" they were required to keep people informed on the progress of their investigation into the deformities.

Tietge and Gary Ankley were ordered down to North Carolina, to a meeting between officials from the EPA and the NIEHS, to defend the criticism they'd made of the FETAX study in Minnesota. The meeting was also meant to smooth over differences between the two agencies, though if it did so that was not apparent in any outward way. Tietge would only say that he and Ankley felt the correctness of their interpretation of the results had been reaffirmed.

Tietge alerted everyone involved that he would present the EPA's FETAX results at the upcoming annual meeting of the Society of Environmental Toxicology and Chemistry in San Francisco in mid-November— where he was co-chairing a session on the deformities problem that would also include presentations from Stan Sessions, Judy Helgen, Dave Hoppe, and Jim Burkhart. SETAC is the main professional organization of toxicologists, and the mere fact that the deformities problem had earned space in a crowded agenda was significant. But the head-on collision between Tietge and Burkhart proved less exciting than it might have been. Tietge politely reported the EPA's work showing ion imbalances were the cause of the positive FETAX results. Burkhart's unruffled response included a blizzard of new data showing that FETAX testing performed using clean "artificial" water with the same ion profiles as were found in the field did not produce abnormal development. This sounded to me a lot like Burkhart was saying that there was in fact *something* in the natural water that caused the effect. But the arguments on both sides were so technical and so methodological in nature that I wasn't sure any of the two hundred or so people in the audience understood that these two guys were fighting with each other.

At least one researcher I spoke with couldn't understand what all the fuss was about. Bruce Blumberg e-mailed me to say that it didn't make any sense to him to argue over ion imbalances or anything else you could manipulate in the water samples to alter the results in FETAX. Blumberg thought it too obvious to even require comment. "What counts," Blumberg said, "are the relevant ion levels in the field."

In late February of 1998, Dennis and Rhonda Bock got a letter from the MPCA informing them that the agency and their research partners had further evaluated toxicology data from some three dozen wells that were located near sites where frog deformities had been reported, including theirs. Without elaboration, the letter stated that these tests on the Bocks' well water were *negative*. Apparently, their water was now all right again.

The same conclusion was eventually reached on virtually all of the wells examined, and it was clear that it would be hard to link the one or two wells that remained suspicious to the deformities problem. None of this was announced publicly.

I called Jim Burkhart and asked him if subsequent testing had led them to a different conclusion about their well-water findings. He said that wasn't what happened. "The wells repeat," he said, meaning that in follow-up FETAX testing on well-water samples they got the same results over and over again. The actual numbers of abnormalities in the assay never changed. But the results fell into a gray area that was not conclusive—and always had. This was essentially the problem Joe Tietge had outlined. The test had a subjective component. Burkhart and the people at Stover had "rescored" and "reinterpreted" the significance of these numbers and now believed that the results they formerly considered positive could just as easily be changed to negative. So that's what they did. Whether this was a purely impartial exercise or the researchers were simply eager to extricate themselves from the mess they'd gotten into over well water is hard to say. But when the investigators backed away from their earlier pronouncements, they went quietly, on tiptoe.

In the fall of 1999, two years after the ill-fated FETAX announcement, I was talking with a hydrologist from the MPCA named Joe Magner. Magner had worked on CWB and was responsible for installing the shallow groundwater wells from which the original alarming samples had come. Magner told me the positive FETAX results on groundwater had raised the level of concern about that site substantially. But that might not have been the case, he conceded, had the agency really understood the hydrology there. Everyone assumed that groundwater was subterranean in origin, and that it likely was a source of inflow to the lake itself. Thus a contaminant in the lake might well be coming from underground—the very nightmare scenario first imagined by Judy Helgen and Bob McKinnell in their work at the excavated pond on the Ney farm.

But Magner said that in the time since the first FETAX studies they'd learned that it was the other way around. The so-called groundwater they'd collected near the shore was actually surface seepage that came from the lake. It had seemed to have a different origin because it was at a different elevation than the lake, but the MPCA hadn't taken into account that the lake level often went up and down seasonally.

"The first time I saw CWB I thought to myself it was just one big bowl of rainwater," Magner said. Although there is a small spring on the north side of the lake, Magner's initial hunch turned out to be essentially right, he told me. Whatever was in the groundwater wells and the sediments at CWB had flowed there from the lake—a fact that made it very unlikely indeed that there would be any related contaminants in the Bocks' well. Various dating techniques that allow hydrologists to determine the "age" of well water in Minnesota indicate that it typically takes years and often decades for surface water to reach deep aquifers.

"I think we were on a wild goose chase at CWB," Magner said.

———

Joe Tietge took no particular pleasure in debunking the FETAX findings, but he was genuinely excited about the surprising results of the methoprene experiment the EPA had worked on all summer in Duluth. It hadn't turned out the way anybody ever dreamed it would.

They'd begun testing the insecticide in five concentrations in early May. Only the pure compound—the active ingredient—was administered. This was standard procedure, as any commercial formulation would contain methoprene plus various carrying compounds and other supposed inerts that might complicate the results. The highest concentration was the maximum amount that was soluble in water—around 500 parts per billion, or roughly 125 times as much methoprene as is needed to control mosquitoes, one of its most common applications. The other concentrations were progressively weaker, down to less than 10 parts per billion.

The test organism was *Rana pipiens*, and the exposure commenced within minutes of the eggs being laid and continued on through successive developmental stages all summer long. The tadpoles were in shallow glass dishes arranged in rows on a long, stainless-steel table in a quiet, windowless lab. The test solution was replenished every two days—a procedure called static renewal, in which the dose is periodically restored as the compound breaks down or is metabolized by the organisms. There were of course also control specimens undergoing the same regimen, only without exposure to methoprene.

At the Duluth meeting, Gary Ankley had talked about the possible role of ultraviolet light in elevating toxic responses in aquatic systems. It was known that sunlight could, in certain circumstances, "photoactivate" chemical compounds that might not otherwise be toxic. With respect to methoprene there was already evidence that UV accelerated its breakdown into the suspect compound methoprene acid. So half the experiment was run under periodic ultraviolet light to simulate sunshine during daytime hours.

Dramatic effects appeared right away. When I visited Tietge early that summer he took me down to the lab and switched on a computer. Presently a series of images came up on the screen. They looked like gray lumps of clay.

"What are those?" I asked

"Well," said Tietge, "they're the tadpoles that got the high concentration. They didn't live at all. One hundred percent mortality. They're just blobs, as you can see." Tietge said he was now just waiting to observe the effects of the lower concentrations, which he expected would be sequentially less severe developmental disruptions, perhaps involving discrete organ systems rather than the whole animal.

He waited a long time, until Labor Day. Nothing happened. Tietge and his colleagues examined subsets of tadpoles regularly, all the way through forelimb emergence. None of the other tadpoles seemed to be affected in any way by the lower doses of methoprene.

"And this was true for both the ones under UV and those that weren't?" I asked.

"Right," Tietge said. "No methoprene effect."

He seemed to be picking his words haltingly.

"However, there was a UV effect," he said after a pause.

"What was the UV effect?" I asked, looking casually at my notes and not quite registering what he was saying.

"Reductions," Tietge said. "Limb reductions."

I sat up straight.

"OK," I said. "What limb reductions?"

"Hind limbs only," Tietge said. "Mostly bilateral, many symmetric. So if the third and fourth digits, say, were missing then they would usually be missing from both sides."

"So you found missing digits."

"Yes," he said. "Also other things."

"And you didn't find these deformities in animals that were not exposed to UV?" I asked, trying to make my voice sound normal.

"None," Tietge said.

We went down to the lab and Tietge booted up his computer to show me pictures of the test frogs. The range of limb reduction appeared to run from missing toes to completely missing legs—with partial limbs of every conceivable length in between. Apart from the fact that both sides of the frogs seemed to be affected equally, the truncations looked to me like many I had seen in the field. Tietge explained that the truncations seemed to be dose-dependent; tadpoles removed from the experiment retained more of their limbs than did the ones that endured longer exposures.

"It appears that the methoprene experiment has morphed into a UV experiment," I said. "What do you make of it?"

"That's the way it goes," Tietge said with a smile. "It's good for your ego when it happens."

I had a million questions. How much UV were the tadpoles getting? Less than they would on a normal summer day, Tietge said. But wasn't their exposure constant since there was no hiding place, no vegetation, in the test dishes? Tietge said that was true. The tadpoles had no place to hide and were in water that was completely transparent. Was the artificial light exactly like sunlight? No, he said, it was a more limited spectrum, and that could be important.

I asked if they had any idea how ultraviolet radiation could cause a leg to stop developing.

"It's an important question," Tietge said. "As far as we know, this is the first demonstration that UV can have an effect on limb development in frogs. There seems to be a time dependence here, because we got different results as we moved organisms out of the experiment in different time frames. Anyway, the limb buds, if you look at them, are just little white colorless protuberances as they emerge. Retinoids absorb UV. This is just a working hypothesis, but it seems possible that the retinoid concentration in the limb could be eliminated. But I'm not so narrow-minded to think that the effect is completely retinoid related. There are other chemicals in the developing limb that could be destroyed because there is no pigmentation to block UV."

"So you think UV might knock out retinoic acid?" I asked.

"Maybe," Tietge answered.

Tietge said the limb reductions in the experiment were in fact bilateral—that is, they occurred on both sides of the animal—in about 70 percent of the tadpoles. Roughly half of these dual truncations were identical on both sides. This wasn't exactly consistent with the field data, where missing limbs were more often reported only on one side of the frog. It occurred to me that a frog that was missing both hind legs—like the frogs Bob McKinnell encountered at Audre Kramer's—wouldn't live long after metamorphosis, so this particular phenotype, even if it occurred in nature, might not be found very often. I asked Tietge about this apparent discrepancy—plus the fact that the sun shines everywhere but deformed frogs had not been found everywhere.

"Toxicologically, this is a significant finding," he said. "What it means for the situation in the field I don't know. It's true that the sun shines everywhere, but it's also true that there's a concern right now about increasing UV. So I'm not sure how this applies. One thing I do know is that some of what we've seen looks very close to what's been seen in the field and this seems to add to the weight of evidence that we're probably looking at a more complicated situation. Not a simple one. It has been very convenient to lump these observations into one phenomenon. But we're probably looking at some deformities that are parasite-mediated, some that are chemical-mediated, some that are UV-mediated, and some that are combinations thereof."

Tietge presented this wholly unanticipated result—along with the EPA's conviction that methoprene was now off the hook as a cause of the deformities—at the SETAC conference in San Francisco during the same talk he gave on his FETAX data. Shortly afterward, I exchanged a series of e-mails with Dave Gardiner and Sue Bryant about the UV effect—for which they proposed an alternative explanation, one that would soon be generally accepted.

Gardiner and Bryant thought the truncations most likely were caused by direct UV damage to the skin at the tip of the limb bud. This outermost bulge of skin, which is the first part of the limb bud to form, is known as the apical epidermis. Its role in the progression of limb development is critical. Many experiments have shown that removing the apical epidermis at any point during limb outgrowth causes development to cease and a truncation results. Gardiner and Bryant believed the UV in Tietge's experiment in some way disabled the apical epidermis, rendering it nonfunctional. "We think the bottom line is that Joe is giving the animals a sunburn," Gardiner wrote.

This hypothesis in no way diminished the potential significance of the UV result. Frogs and sunlight have coexisted for millions of years. Yet, like Andy Blaustein in the mountains of Oregon, Tietge had stumbled onto evidence that ordinary sunshine could alter the course of amphibian development. Understanding the mechanism of disruption in a way seemed only to reinforce the significance of the discovery. Frogs that were apt to develop without hind legs would be at a profound disadvantage in evolutionary terms—they should have been weeded out long ago if the cause were ordinary sunlight. Either that or ordinary sunlight was changing, in which case the weeding out might only be starting.

Ever since the Oceanside meeting, Gardiner and Bryant had been spending a lot of time with Dave Hoppe's specimens from CWB. Gardiner wrote that there was an important difference between the blunt truncations Tietge obtained via UV exposure and some of the morphologies they saw in Hoppe's frogs. In the UV-treated limbs, the leg grew out normally for some distance and then simply stopped. But Hoppe's frogs often featured legs that grew out from the body for the correct distance but showed a "progressive loss of structure" as they did so. In other words, Gardiner wrote in an e-mail, some CWB frogs had legs that were tapered rather than truncated.

Gardiner said they were struggling to interpret the various limb structures in the Minnesota frogs, many of which he said were very complex. But he teasingly hinted at something they'd found—an important morphological marker, in a significant number of the abnormal limbs. "Something Sue and I didn't at first think we'd ever seen before," he said. But then, after several late nights of discussion and argument, Gardiner said they'd gone back to the scientific literature one more time and there it was after all. Gardiner wouldn't tell me yet what the marker was. But he did say where it had turned up before: in experiments involving retinoid-treated limbs.

Gardiner had been busy all fall. In the coming year his work on the deformities problem and the attention it received by the federal government would profoundly reshape the investigation.

Immediately after the Oceanside meeting, Gardiner and Bruce Blumberg had contacted Dave Hoppe and begun putting together an experiment to run at CWB using what was left of their $5,000. They also got Ron Evans, Blumberg's boss at the Salk Institute to grant them free use of Blumberg's retinoid receptor assay to screen water samples. In October Gardiner e-mailed me to say he would be coming to Minnesota as soon as the test "apparatus" was ready. Blumberg, he said, had built a special piece of equipment to collect a highly concentrated water sample from CWB. If it would be all right, Gardiner said, he wanted to ship this equipment to me, since part of it consisted of several large tubes that he might have trouble checking with his baggage as they looked "quite a bit like small missiles."

It was the first week of November and when the packages arrived, followed a day later by Gardiner himself. We left the next morning for Brainerd. The day was somber and chilly. Patches of snow lay in the ditches by the road. It seemed a strange time to be working on frogs.

Dave Hoppe and the Bocks were waiting for us when we arrived and after a few brief introductions Gardiner began setting up his equipment on the shoreline next to the dock. I had been worried the lake might be frozen this time of year, but CWB rippled crisply in the icy breeze. A few knots of migrating ducks paddled in circles offshore and a large flock of Canada geese periodically lifted off from the lake's slate-colored surface to wheel in the sky, testing the air to see if winter was in it.

Gardiner's equipment consisted of four long, clear, double-walled plastic cylinders that were connected with tubing at the bottom. The inner column of each was filled with tiny resin beads that looked like fine white sand. Blumberg had rigged up a fish aquarium pump to draw water from the lake and push it through the four tubes. Organic compounds in the water would stick to the resin beads. The longer the pump ran, the more water would pass through the columns and the more densely concentrated the organic compounds would become. When Blumberg received the columns back in La Jolla he would wash the collected compounds from the beads and assay for any that activated the retinoic acid receptors. If there was a chemical in CWB that mimicked retinoic acid—and it was still present at this time of year—it would show up.

Gardiner and Hoppe spent a few minutes lying on the dock, their hands freezing in the water as they got the intake hose in place and primed the pump. The columns filled with water and the experiment began humming along. Rhonda Bock made lunch for everyone. Hoppe had to get home. Gardiner and I spent the afternoon walking with Brandon and Troy Bock in the woods while the pump ran. It was the day before the opening of deer season, and I pointed out the tree stands nestled among the barren branches of the forest to Gardiner, who was having a splendid time.

"This is great," he kept saying. Gardiner had once worked in marine biology, but since getting into limb development it had been years since he'd been in the field. Gardiner was fine company—funny and enthusiastic. He seemed to see the deformities issue not so much as a freestanding problem to solve and be done with but as small part of a continuum of evolving questions that were out there for science to explore. He talked about science in the way other people might talk about the current football season or the ups and downs of the stock market—like it was an unfolding drama, one paced by exciting gains and soul-searing setbacks. Like a number of university-based scientists I spoke with, Gardiner was skeptical about the quality of the research that was being done within government agencies. The best scientists usually went into academia or to high-paying industry jobs, not into the public sector, he said. Not that academics were perfect by any means—Gardiner had had his share of run-ins and career reversals on campus. He said he liked his research position at Irvine much better than being a professor, which he hadn't cared for much at all. "You know what it is?" he said. "You're dealing with people, your peers, who spent their formative years sitting in the front row of class in thick glasses, with pocket protectors, answering all the questions right. The people nobody else could stand. Then they go to college and they go to graduate school. They become professors and now they get some power. And then they become a huge pain in the ass. Authority corrupts them."

Gardiner was laughing all through this.

"So science on the inside is revenge of the nerds?" I asked.

"You got it," Gardiner said.

It grew dark around 5 P.M. and Gardiner shut down the pump and began repacking the columns. He also filled some good-sized jars with water from the lake. "I don't know exactly what these are for," he said as he taped the lids tightly to prevent leakage on the way back to California. "Souvenirs, I guess."

––––––––––––––––

Bruce Blumberg wouldn't get to work on the resin columns until early in 1998. He e-mailed me that he was astonished by the large number of organic compounds that were present in CWB water. This was the same thing Jim Burkhart had told me—it seemed to be a characteristic of wetlands that produced frog deformities. Although NIEHS had yet to identify a single chemical agent common to all the deformities sites that might account for the abnormalities, they did find that there were always more chemicals in affected ponds than in reference sites.

Initially, Blumberg didn't detect any retinoid activity in the samples. But in subsequent runs he did find that something in the water activated a retinoic acid receptor.

In February he wrote me that something seemed to be present at low levels in the water that could be consistent with developmental disturbances from a retinoid. Blumberg expected that whatever the compound was, it would be hard to identify. But the missing piece of the puzzle was there. Blumberg was confident of that much. He said the one thing he still needed to check was the resin itself—just to make sure that it was truly inert.

Blumberg was shocked when he discovered that the resin beads he'd used in the columns *did* activate one of the retinoic acid receptors—a surprise that rendered the whole experiment invalid. The only remaining hope was that there was enough water in the jars Gardiner had filled at CWB to detect a retinoid in the lake if there was one.

There was enough water. And there was a retinoid in the lake.

––––––––––––––––

In December, Jim Burkhart held a workshop on the deformities problem at the NIEHS headquarters in Research Triangle Park. Most of the active participants in the investigation were there, although Stan Sessions and Joe

Tietge were notable absentees. Dave Gardiner had a plane connection through Minneapolis and we ended up on the same flight to North Carolina.

Gardiner was full of news. He and Sue Bryant had been thinking about some of the questions I'd had for them at Oceanside. Everything still pointed strongly to some sort of retinoid involvement in the deformities. And the fact that deformities often occurred on only one side of the frog no longer appeared so strange, he said. Asymmetry was actually a hallmark of developmental defects, and the reason was that development itself is generally not symmetrical—which was plain to see if you looked at the arrangement of internal organs and other important differences between the right and left sides of a vertebrate organism.

But what Gardiner was most excited about was the marker he and Bryant had discovered in the Minnesota frogs. He now explained that it was a unique bending of the leg bone, such that the two ends were pushed toward each other, the midsection forming an apex in between. "We haven't decided what to call it yet," Gardiner said. "They're like pyramids of bone, but not quite."

"A bony triangle?" I asked

"Sure," he said, "why not. The point is, these bony triangles have shown up in a lot of the retinoid experiments in the past."

The plane was still climbing over the brown fields and frozen lakes of the upper Midwest as Gardiner talked. It was late afternoon. The sun streamed in through the starboard windows. Gardiner said he would make the case for retinoids at the meeting the following day—even though Blumberg would not begin assaying the water from CWB until the following month. Now that they had morphological evidence of a retinoid exposure it seemed more probable than ever that this was a plausible cause of the deformities.

The first speaker the next morning was Mike Lannoo, who had also been at work on the deformities problem, X-raying abnormal frogs from Minnesota and other places. He had told me that he too had found a morphological marker—one he believed argued that there were different causes for the deformities in different locations. Lannoo had repeatedly seen in both truncated limbs and extra limbs an odd softening and expansion of the leg bone—a ballooning—that was most often located at the terminal end of the bone, but also within bone segments. This "spongiform bone," as Lannoo called it, did not occur everywhere deformities were being reported.

Since Lannoo felt he needed to work on the spongiform observation more before he could make it even a working hypothesis, he instead used the occasion of Burkhart's meeting to urge everyone else to exercise similar restraint. The deformities investigation, Lannoo said, was out of control.

"Have you ever seen a scientific issue debated so contentiously?" he asked. There was nervous laughter. Lannoo said everybody ought to remember a few basic concepts as they proceeded in the future. Be wary of "model" species, Lannoo said. Generalizing from a response in *Xenopus* in the lab to what happens to a native frog in the wild was, he said, "a stretch." Lannoo also asked for caution in overstating the importance of frogs as sentinel species. "There are many indications that we live in an environment that is in decline," he said. "But can we really say that frogs are a good indicator of how those conditions may affect humans?"

Lannoo lamented the lack of peer-reviewed results in the investigation and the abundance of "unilateral declarations" about potential causes and the "shoehorning of facts" to fit hypotheses. The objective should be the reverse—to search for data that would refute a hypothesis. A hypothesis refuted, Lannoo added, is no discredit to its author.

"Hypotheses are meant to be tested," Lannoo said, "not argued over. If you want to make a career of arguing hypotheses, then become a theologian."

Gardiner and Blumberg spent the winter working through the connection between the bony triangles in the frogs from CWB and the fact that a retinoid had been found in the water there. In March, they presented these tentative findings at a meeting of the Declining Amphibian Populations Task Force in Milwaukee. They stressed that this was only preliminary work from a single site—but the fact that two separate pieces of evidence pointed in the same direction demanded further work on retinoids. The stakes, they said, were very high.

"Bioactive retinoids in water are a definite public health risk," said Blumberg. "Retinoids cause developmental deformities in every vertebrate species that's been tested, from primitive fish to humans."

The retinoid, or retinoidlike substance, in the Minnesota water could be a pesticide or a derivative of one, Blumberg said. But he cautioned that it was also possible that it was a natural compound produced by plants or microorganisms in the lake. Even it if were natural in origin, he said, that "just means there's nobody to blame."

Gardiner said they would continue to assay more samples from Minnesota, including water from reference sites, in an effort to identify the compound that activated the receptor. But the real work would be to induce comparable deformities using the suspect chemical once it was found. "We're going to have to produce a frog with extra legs in the lab," he said.

Gardiner couldn't have known—nobody in the investigation could have known—that somebody else was about to do exactly that.

That same spring, Gardiner participated in two high-level briefings in Washington in the office of Interior Secretary Bruce Babbitt, who was interested in the population declines and deformities that were being reported in frogs. Two of the participants in the briefings were Dave Wake, from Berkeley, and George Rabb, director of the Brookefield Zoo in Chicago and a member of the international Species Survival Commission. Wake and Rabb had been instrumental in setting up the Declining Amphibian Populations Task Force. DAPTF had been monitoring the status of amphibians for the better part of a decade; now Wake and Rabb wanted a better-funded, more centrally controlled organization that could send teams of field investigators into hot spots around the world to assess amphibian problems the instant they were identified. There was an obvious logic in marrying the declining population issue to the deformities problem, since the high visibility of the deformities investigation offered the prospect of greater support for frog research on Capitol Hill.

Babbitt was sufficiently impressed with the presentations to begin organizing a multiagency effort to address frog issues. This in turn led to a series of three meetings sponsored by the National Science Foundation, beginning in the spring of 1998. The last of these meetings, which would take place at the San Diego Zoo in November, would focus on developmental aspects of the deformities problem. Dave Gardiner was selected to organize the workshop.

Gardiner was ecstatic. Here at last was a chance to focus the discussion where he, Sue Bryant, Ken Muneoka, and Bruce Blumberg believed it should have been all along—on the biochemical pathways that had to be affected by some environmental factor that was producing the deformities. This would be the meeting where arguments over field protocols and toxicological methodologies—which bored the developmental biologists to distraction—could be avoided.

But in October of 1998, Gardiner called me sounding dejected. Someone—he thought it was probably Dave Wake—was interfering with the agenda for the meeting and insisting he invite two people to make presentations who weren't on the original list of invitees. One was Stan Sessions.

"I can't stand it," Gardiner said. Sessions's public declaration that parasites·were the cause of the deformities outbreaks had continued to rankle Gardiner and his colleagues. He didn't believe the theory and he didn't like the attention Sessions managed to get for it. But senior officials at the NSF told him to invite Sessions and let him speak. Period.

"Who's the other person?" I asked.

"I don't know," Gardiner said. "Never heard of him. Somebody named Johnson. Do you have any idea who he is?"

"Not Pieter Johnson," I said.

"Yeah," Gardiner said, "that's him."

I told Gardiner I did know who Johnson was. But I couldn't believe anybody at the NSF did.

———————

I'd met Johnson the previous fall, at the SETAC meeting in San Francisco. He was a twenty-two-year-old senior at Stanford, an undergraduate. I'd run into him in a coffee shop in the hotel where the meeting was being held. He was sharing a table with Stan Sessions. When I asked Johnson about his interest in deformed frogs he informed me he'd been working on several ponds north of San Francisco where abnormal Pacific tree frogs had been found. He even had some pictures with him, and he showed them to me. The deformities looked a bit like the ones at CWB, although the tree frogs were very small and seemed to have extra legs as their predominant malformation. I asked Johnson if he had any ideas about the cause of the deformities. He said he was working on parasites.

It took me several tries to get ahold of Johnson after I spoke with Gardiner. Johnson had graduated and was spending a lot of time in the field. Finally I got through. I told him I'd heard he was coming to present his findings at San Diego the following month and was curious what kind of data he had.

Johnson remembered me and was more than happy to talk. He had continued to work on parasites, he said, and had some interesting results that integrated field observations with a laboratory experiment. He and a couple of undergraduate pals he'd recruited to help him had identified two trematode parasites that seemed to be closely associated with deformities in four ponds that had been reported to biologists at Stanford back in 1994. Nobody else had been interested in working on the deformities, but Johnson and his friends identified the snail that was the first intermediate host for the parasites. Eventually they determined that one of the parasites, of the genus *Ribeiroia*, was always present in deformed frogs and that it tended to infect a localized region at the base of the hind limbs. Johnson then started bringing snails into the lab and learned how to harvest the cercarial parasites as they were shed into the water. Finally, he'd conducted a live assay using Pacific tree frog tadpoles that he had exposed to *Ribeiroia*.

"What did you get?" I asked.

"Everything."

"It's the full Monty!" Stan Sessions chortled when I spoke with him shortly after getting off the phone with Johnson. The experiment with *Ribeiroia* had produced virtually all of the types of deformities seen in the field, in California as well as Minnesota. Missing limbs. Extra limbs. Skin webbings. Even, Sessions claimed, "bony triangles." Sessions had been in close contact with Johnson all through the winter and was overjoyed at these results—which appeared to vindicate the parasite theory convincingly, at least in the case of Pacific tree frogs in California. The fact that an agent identified with deformities in the field caused the same deformities in the lab under controlled conditions made the findings especially compelling. Naturally, Sessions thought the implications were very broad. Johnson's major breakthrough, Sessions said, had been in identifying the right parasite. Sessions told me he "hadn't known squat" about trematodes back when he started on the deformities problem at Aptos and that he'd wasted a lot of time working with *Manodistomum*. Now that they knew which parasite they were dealing with, experiments could commence with other species of frogs. Sessions said he already knew that *Ribeiroia* was in Oregon, New York, and California—and because its definitive host was a bird it could travel.

"I call it 'the bad boy,'" Sessions said. Together, he and Johnson would make their case in San Diego. Sessions seemed to think the role of parasites in the deformities outbreaks would shortly be beyond dispute.

"I'm more convinced than ever that multilegged frogs are caused by parasites," he said.

The Green Mountain

STAN SESSIONS ONCE ASKED ME WHY I THOUGHT HE SEEMED TO rub everyone the wrong way. I told him I wasn't entirely sure, but I thought it was at least partly due to his forceful speaking style. "Maybe they're jealous," I said. "People dislike being convinced that something they don't agree with might be true." Sessions shook his head. He told me that he often felt angry over speculation about the deformities. Ill-conceived ideas seemed to strike him as fundamentally dishonest. "Any attempt to mislead," he said, "always infuriates me." This was, of course, exactly the kind of loaded statement that endeared him to no one in the deformities investigation.

But the National Science Foundation meeting at the San Diego Zoo in November of 1998 was not the triumphant victory party for the parasite theory Sessions had hoped for—although Pieter Johnson's findings on *Ribeiroia* made a big impression on everyone there. During the closing session, as a consensus statement was being hammered out, Sessions had made an impassioned argument that parasites should at least now be considered the most probable cause of the deformities—the prime suspect against which all other proposed causes had to be measured. But the group—fifty or so esteemed developmental biologists, plus the familiar cadre of field investigators and toxicologists at work on the deformities problem—declined to elevate any one hypothesis over another. Parasites were still on the table. So were chemicals. So was ultraviolet light.

"Maybe it is the way I present things," a disappointed Sessions said to me afterward.

For his part, Pieter Johnson was brilliant—his presentation sparked the most intense discussion of any at the conference. It came at the end of the first day, as the setting sun lit up the lush, steep hillsides of the zoo grounds and cast long shadows through the green shutters of the conference room—an airy, tree house–like structure that was made just dark enough for Johnson's slides to be projected on the screen next to the lectern. The mood among the scientists was relaxed, and surprisingly upbeat at the close of what had been a long day of talks on limb development and environmental factors that might influence the process. Quite a few of the participants had spent the midday break sauntering through the zoo, relaxing in front of the hippo tank, or buying souvenirs in the gift shop. Only a handful of people in the room had any idea what Johnson was about to show them.

Johnson got up to speak. He was very young—fit and energetic, with close-cropped hair that made him look like a junior military officer. He struck me as intensely bright and confident, although I would later hear other people describe him as cocky and arrogant. I thought he struck exactly the right balance between self-assurance and a proper respect for his audience. Either way, he managed a riveting account of his experiment to a crowd of skeptical scientists, most of whom were twenty or thirty years his senior.

The focus of his work, Johnson said, had been four ponds among the thirty-five he'd surveyed in Santa Clara County, where he'd found deformities in several species of frogs and some salamanders. In all, he said he'd examined as many as 15,000 amphibians. Pacific tree frogs, *Hyla regilla*, bore the brunt of the malformations, with between 25 percent and 45 percent of the frogs affected. Extra limbs were the most common deformity, accounting for about half the malformations observed, but other frequently observed abnormalities included missing limbs, twisted or permanently extended limbs, and cutaneous fusions. Johnson also found bone abnormalities that he described as bony triangles.

Johnson said the first important clue as to the cause of the deformities was finding that the four affected ponds were the only ones that were home to a species of snail that harbored *Ribeiroia*. Dissections of abnormal frogs showed the same parasite encysted in its metacercarial resting stage in the pelvic region of the deformed animals. In the lab, Johnson was able to collect *Ribeiroia* and infect developing *Hyla regilla* tadpoles at specific rates ranging from sixteen parasites per frog to three times that. *Ribeiroia*, which located itself at the base of the limb, never out along the limb, induced deformities at all doses. Johnson said he believed a single metacercarial cyst would be enough to cause an abnormality.

What made the experiment convincing to him, Johnson said, was that the infection rates were "realistic," the cercariae were allowed to infect "at will," and the parasites used were the same ones associated with the deformities in the field. All of this—plus the fact that the test organism was the same native species affected in the wild—created a "tight agreement" between the lab results and the field data. Johnson also noted that the deformities obtained experimentally constituted about "95 percent of the abnormalities seen in the field" and pretty closely matched the rates of specific types of malformations. In other words, the deformed frogs created in the lab were essentially indistinguishable from the ones collected in the wild.

Earlier that year I'd gotten in trouble with Judy Helgen and Joe Tietge when I had reported on the Gardiner and Blumberg retinoid results for the *Washington Post* and compared the $5,000 they'd spent to the hundreds of thousands of dollars being spent by various government agencies investigating the deformities. Helgen and Tietge thought the comparison unfair and inaccurate. When I talked with Johnson the day after his presentation he said the total cost of his experiment had been "less than $500."

Johnson had been helped in the field and in the lab by a couple of other students, including his boyhood friend Kevin Lunde, whose family lived just a couple blocks away form Johnson's in the quiet, prosperous suburb of Claremont, just east of Los Angeles. Johnson and Lunde recalled for me how they'd kept expenses down. When they needed 250 small plastic containers for rearing tadpoles they'd gone to a local Wal-Mart, explained they were students working on a science project, and asked if the store would donate the containers. The store manager said the best they could do was give them fifty. Johnson and Lunde accepted the offer gratefully—and then repeated the same plea at four more Wal-

Marts. When they needed clean water for the experiment they stole it from water coolers on the Stanford campus—sneaking through buildings late at night and never taking a large quantity from any one location to avoid detection.

"The spring water was really great," Johnson said, "because they refill it really fast when it gets low."

———————

While Johnson's presentation earned him high marks, the experiment itself got mixed reviews at the San Diego meeting. Andy Blaustein, one of the relatively few scientists familiar with the overall investigation who had maintained all along that parasites were a viable possibility, thought it was highly important and persuasive. Mike Lannoo liked the simplicity of the procedure and the clear results. But Dave Gardiner was frustrated by a lack of detail about exactly what happened and at what stage. Sue Bryant told me that, after years of working with students, she'd learned to be cautious about results obtained in experiments by people "who don't have a track record."

"I mean, you just don't know they did it right," Bryant said. "You need to know who helped them. You need follow-up studies to more closely stage the results. You need to have confidence that they did the controls properly. It would be really hard for me or anybody else to change their thinking based on this one experiment by undergraduates. That sounds harsh, perhaps, but it's dangerous to assume that inexperienced people will get it right."

What puzzled a number of the scientists was that Johnson seemed to get deformities throughout the limb that were caused by a parasite cyst at the base of the limb. How was it possible that a leg could grow out from the body normally for some distance and then be deformed farther out if the cause was located some distance away? What sort of "signal" could a cyst generate that would travel across a normal segment of a leg to disrupt development farther out? Gardiner and Bryant thought it flew in the face of basic developmental theory. There was a good deal of discussion too about whether the parasites caused deformities by mechanically disturbing the limb or via some sort of chemical signal—either by generating one themselves or by initiating one in the host.

Ken Muneoka said later that these two possibilities could reasonably be construed as essentially the same thing. Mechanical disruption of the developing limb caused positional confrontations among cells that were resolved via intercalation. Chemical disruptions did more or less the same by "resetting" the positional values of cells in the limb. Thus any possible involvement of parasites in some kinds of deformities did not rule out chemicals as a cause of similar deformities in other places—and vice versa.

Stan Sessions told me he thought it was time for everyone to now admit that the whole deformities problem had been "overblown." You could separate the deformities into two categories, he said. There were frogs that exhibited a range of abnormalities, including extra legs, that made up one category. These, he said, were caused by parasites. Then there was the second—and, he admitted, more common—category in which the deformities consisted solely of missing or partial limbs. Sessions said these deformities were much harder to interpret, but his own observations still suggested trauma as the most likely explanation. He'd seen it in the field and in the lab, where he said he'd observed tadpoles attacking each other's legs. But this seemed to me to only add further confirmation to Martin Ouellet's observation that trauma was easy to detect where it occurred.

Sessions also said he had looked at frogs with missing legs, mainly from Vermont, and had not found parasites associated with the deformities. Meanwhile, he'd seen *Ribeiroia* in a number of frog species that had been sent to him from around the country. He told me he'd seen the cysts by clearing and staining frog.

One person at the meeting who took an intense interest in Johnson's results was Dan Sutherland, a parasitologist from the University of Wisconsin, La Crosse, which is on the other side of the Mississippi River from southern Minnesota. Sutherland said he'd examined a large sample of frogs with missing legs from a site near La Crosse and could say with absolute certainty that they were not infected with trematode cysts. But he said he would look again and also try to examine more frogs from that part of the country.

———————————

Ken Muneoka thought parasites were at best only a partial explanation for the deformities—a novelty of nature—and that if parasites were your only proposed explanation, you could never be more than a marginal player in the investigation. Scratching his head one evening over dinner with Gardiner and Bryant, Muneoka said if Johnson and his friends were correct, there wasn't anywhere else for them to go with the findings. "If they're right," Muneoka said, "they're out."

This was true, at least to the extent that frog deformities raised potential human health concerns. Although it was possible that parasitism might be increasing because of contamination of surface waters—environmental changes that might have other effects—the reality was that parasites appeared to be primarily a threat to amphibians. As David Gardiner told the *Los Angeles Times*, "I don't think anybody's lying awake at night worrying about parasites."

But the parasite research was far from done. In April of 1999, Pieter Johnson published his findings in the journal *Science*—alongside an article by Stan Sessions that argued retinoids could not cause the deformities seen in the field. This occasioned wide covereage in the lay press, mostly in articles that took their cues from *Science*, leading with the news that the deformities mystery was "solved" and burying the fact that parasites did not explain all the outbreaks. Both Sessions and Johnson said they had tried to convince all the reporters they spoke with not to present parasites as a final answer to the deformities question.

In the summer of 1999, Pieter Johnson and Kevin Lunde made a long cross-country road trip on which they sampled frogs from more than one hundred ponds, including reference sites and places where deformities had been reported to NARCAM and other sources. They traveled through the Pacific Northwest and then east to Minnesota. It was not, Johnson told me, a random survey by any means. They found *Ribeiroia* wherever they found deformities, mostly in *Hyla regilla*, but also in several ranid species of frogs and one population of salamanders. Johnson said he was convinced that *Ribeiroia* was the cause of the malformations wherever "the system" was in place—meaning the snail-parasite-frog deformities complex.

In August 1999, I spoke with Dan Sutherland, who was quietly trying to work with both Johnson and the Minnesota researchers. He'd finally gotten a chance to dissect frogs that Dave Hoppe sent him from CWB, plus some from the Ney farm that he'd gotten from the MPCA after a long haggle with Judy Helgen. He told me he'd found *Ribeiroia* at both sites.

Sutherland told me the *Ribeiroia* infection rate in the mink frogs he'd examined from CWB was heavy. The cysts were located in the pelvic region, close to the bases of the limbs. The infections were less pronounced in the Ney frogs, and Sutherland said he was particularly puzzled by one specimen that contained only a single cyst that was located on the side opposite a deformity. All the cysts were found immediately under the skin. This was perplexing news—Sutherland said he thought it was likely that skinning frogs prior to clearing and staining would remove most of the cysts. Pieter Johnson had told me that he considered clearing and staining a poor way to evaluate for parasites, that even if the cysts remained after skinning they were rendered almost invisible by the clearing process. Stan Sessions continued to insist that clearing and staining was essential. He found trematode cysts that way all the time. When I asked him why the cysts he'd originally examined from Aptos appeared so dark and obvious to the naked eye, he said it was a total mystery. It was possible, he said, that those had been a different kind of parasite. Johnson agreed that the Aptos specimens were highly unusual. "That's the only time I ever saw cysts show up like that in cleared-and-stained frogs," Johnson said. But he categorically rejected the idea that any other species of trematode caused deformities. It was *Ribeiroia*, or it wasn't parasites.

Johnson asked me if I was surprised at the discovery of *Ribeiroia* at CWB. I said I was and I wasn't. I'd thought the pictures he'd shown me of the frogs in Santa Clara County back in 1997 had looked a little like the frogs at CWB, so finding an apparent link made a certain amount of sense. On the other hand, CWB was arguably the most intensely studied deformities spot in the world. Nobody else had ever found any connection between the deformities there and parasites—including Dave Hoppe, who'd just recently cleared and stained a large sample of mink frogs and found nothing.

When I called Whittier's Stephen Goldberg and Rebecca Cole of the National Wildlife Health Center both again confirmed that they had not seen *Ribeiroia* in Minnesota frogs previously. Goldberg was particularly adamant. "It's a very distinctive parasite," he said. "I didn't find it." He said he couldn't recall which sites he'd gotten frogs from—or even if he'd been informed—but from his description of the types of deformities he'd seen it sounded as if at least some of them must have come from CWB. Goldberg also said he did not consider Johnson's experiment conclusive at all.

"Look," he said. "I like pepperoni pizza. But if you sit me on a chair and force-feed me pepperoni pizza continuously I'll get sick. It's the same thing with parasites. If you put them in a confined space with a tadpole and they target the limb buds, then maybe you'll get leg deformities. That doesn't mean that actually happens in the wild."

Johnson and Lunde had hoped to nail down the parasite connection in Minnesota when they arrived in late August 1999, but their field efforts yielded only ambiguous results, Part of the problem was that none of the Minnesota researchers wanted them anywhere near their sites. Judy Helgen relented and allowed them to visit the Ney farm, but would only permit them to take a handful of frogs. In a coincidence of cosmic proportions, Johnson and Lunde stopped for lunch at an Arby's near Brainerd while en route to another site and struck up a conversation with a young woman behind the counter who turned out to be Jennifer Bock. Johnson didn't know the name, but he quickly surmised that the lake Jennifer described near her home was in fact CWB. He asked politely if he could come out for a look, and Jennifer called her father—who said no. Despite Johnson's later protestations that it was all quite innocent, Hoppe never believed him and resented what he thought was an attempt to sneak onto his research site.

Johnson asked me why people in Minnesota were so uncooperative. I agreed that it was unfortunate—Johnson and Lunde were traveling with a portable lab where they could examine snails and dissect frogs on the spot—but that, at least in Hoppe's case, the outbreak sites were considered highly sensitive. But I also thought another reason might be that Stan Sessions had once again poisoned the investigation and alienated other investigators. In July, Sessions had written letters to Interior Secretary Bruce Babbitt, Minnesota governor Jesse Ventura, and Secretary of Health and Human Services Secretary Donna Shalala—whose department included the NIEHS as one of its agencies—arguing that parasites could explain the "entire range of observed limb deformities" and that research into retinoids and other proposed chemical causes were a waste of time.

This was a sharply different message than the one I got whenever I asked Sessions if there could be explanations for the deformities other than parasites. Chemicals and other causes could not be ruled out, he'd always said. Johnson concurred. "We are not saying that all deformities are caused by parasites or even that all deformities have biotic causes," Johnson said. "If we can find places where the deformities exist and

Ribeiroia is not present, then we know there has to be another factor at work."

Johnson said differences in the composition and frequency of deformities that were seen between their sites in California and other parts of North America could easily be due to variations in the way different species responded to parasite infection. Variable phenologies—annual life cycles—also no doubt played a part. Pacific tree frogs, for example, have a long breeding season that produces cohort after cohort of young frogs, one after the other, so that there are always frogs at sensitive developmental stages available for infection. Frogs in the colder climates of Minnesota, Vermont, and Canada, however, breed in tightly constricted windows of time and produce essentially only single cohorts of young. Johnson said that in their own limited observations of ranid frogs, parasites seemed to cause a greater frequency of truncations and skin webbings.

This was not completely convincing, however. Stan Sessions e-mailed me to report that in unpublished experiments he had conducted with several ranid species, *Ribeiroia* produced a range of deformities that always included extra legs as the largest category of abnormality. The phenology issue was more complex—certainly there were profound variations in the natural history of different frog species that might alter their response or sensitivity to parasite infection. But Johnson's own experiment, conducted in a lab where there was no natural variability at all, had produced the same deformities seen in the field in almost exactly the same frequencies. *Ribeiroia* seemed to produce a signature syndrome that always involved a spectrum of abnormalities in which extra legs topped the list. This went to the heart of Sessions's claim about parasites causing "the entire range" of deformities. It seemed the truth was that there was *always* a range of deformities, never just one. Johnson told me that he had never seen a site with *Ribeiroia*-induced deformities that consisted only of missing or partial limbs. When I spoke with him about this in August 1999, I said I thought it was unclear whether parasites would ultimately explain most of the deformities seen across North America or only a tiny fraction.

"I'd have to say I agree with you," Johnson said.

———————————

In the end, the discovery of *Ribeiroia* at deformities sites in Minnesota, like so many discoveries before it, only resulted in a new set of questions.

Johnson said he was convinced the parasite was the cause of the deformities at CWB. "But I hate saying that without being allowed to go there and see for myself," he told me. Other places in Minnesota were harder to evaluate. One site, which Johnson called "Danube" was located in the west-central part of the state. It was a gravel pit surrounded by soybean fields that had been reported on NARCAM. Johnson and Lunde found about 20 percent of the northern leopard frogs there had missing or truncated legs—but on dissection they couldn't find any *Ribeiroia*, although the did find the host snail in the pond. Equally confusing was the Ney pond. Johnson said the Ney pond was not a clear-cut case of parasitism. "*Ribeiroia* is there," he said, "but at very low numbers. Something else could be going on there."

This thought had occurred to me, too. If parasites belonged to the set of natural causes for amphibian deformities over which some new cause had been superimposed, then there was no reason to insist that only one cause could be acting in a single wetland. Martin Ouellet, for example, was unshaken in his conviction that there were many potential causes for the deformities. In October 1999, Ouellet sent me his long-awaited review of all the known causes previously documented. He found published reports that included a long list of both biotic and abiotic causes for limb abnormalities in amphibians, including: abnormal regeneration after injury; agricultural pesticides; the chemical composition of water; the presence of fish, disease, high tadpole densities, extreme temperatures, hereditary mechanisms, nutritional deficiences, parasites, radiation, ultraviolet light, viruses, metals, and other chemical contaminants. Some of these factors, Ouellet added, could interact with one another in a complex web of causation.

This last observation paralleled something Jim Burkhart had told me on a visit he made to St. Paul earlier in the summer of 1999. Burkhart said he'd come to believe that multiple factors were likely in at least some outbreaks of deformities. He said he thought frogs had become victims of a "convergence of environmental misfortune" and that tried-and-true toxicological appoaches that searched out single factors were likely to fail.

Since the FETAX fiasco in fall 1997, Burkhart and his colleague Doug Fort of the Stover group had set aside their well-water tests and instead focused on surface water and sediment samples from Minnesota. They continued to test them with FETAX, but they added an extended protocol that continued the assay past the standard 96 hours, all the way out through metamorphosis. In October 1999, Burkhart and Fort published two long, extremely detailed papers analyzing the results. They found strong evidence that water and sediments from sites with deformities in the wild were toxic and teratogenic in controlled laboratory tests, impacting rates of growth, tail resorption, and limb development. In longer runs of the assay, *Xenopus* did in fact develop abnormal legs, although the flexures and rigidities observed in the lab did not correspond exactly to the deformities in the field. The tests again did not identify any compound common to all sites that could be directly implicated in the deformities, but Burkhart and Fort did determine that there were high loads of organic contaminants in the deformities sites and that one effect that occurred everywhere was a suppression of thyroid function—a key component of metabolism, and, in frogs, an endocrine-regulated system that is critical in metamorphosis. There was even evidence to support Burkhart's long-held suspicion that there was something peculiar about the water "matrix" in Minnesota. In one startling experiment, when water from a control site in Minnesota was added to a sample from a deformities site, the effects were not reduced by the dilution. Instead, the sample became more toxic.

"Personally, I think *Ribeiroia* is everywhere," Martin Ouellet told me in August 1999. "Snails are everywhere, too. But deformed frogs are not everywhere."

Ouellet said he did not believe metacercarial cysts located under the skin could cause abnormal limbs—but that his real reservation about parasites as a general explanation for deformities continued to be the ironclad correlation he still found in Québec between deformities and agricultural ponds. "Someone would have to explain to me why I have no deformities in control sites where I have more snails and more parasites," he said. "It's clear-cut. Parasites are a natural cause for a pattern of deformities like you see in *Hyla regilla* where you are having many extra legs. But they're not the only cause. Missing legs and shortened legs are more common."

Ouellet sounded exasperated. "So we have to consider again traumatic amputation," he continued. "Right?"

"Yes," I answered.

"So, I say again," he said, his voice rising, *"why do I not see these missing legs in my control sites? Somebody has to explain that!"*

Ouellet told me he had made several large collections of control specimens over the summer at Mount St. Hilaire—as many as 1,500 frogs per collection. At most, he saw a single missing limb in such a sample. Usually, nothing. Meanwhile, at his other study sites, the total sample size had swollen to 45,000 frogs. Nothing had changed. Deformities—mainly missing or reduced limbs—continued to be found at high frequencies in agricultural ponds with pesticide inputs, but not in reference sites. "It's the same pattern every year since 1992," Ouellet said.

In the States, Mike Lannoo also remained convinced that something as yet undiscovered was behind the many reports of missing limbs. I asked him if he ever considered the possibility that the whole deformities issue was, as Stan Sessions put it, "overblown" and all the result of natural, biotic factors. "Never," he said without hesitation.

Lannoo had forged on with his X-ray work, and the pictures—which were properly called radiographs—continued to show spongiform bone morphologies that occurred in some locations, but not others. Lannoo said that this was proof to him that different causes were at work—although he conceded that more work would need to be done to make sure the effect was not something related to which species of frog you looked at. What seemed to Lannoo to be most significant about sprongiform bone was that he did not see it in frogs infected with parasites. In the fall of 1999, Pieter Johnson sent Lannoo a large sample of frogs that included specimens infected with *Ribeiroria* as well as some that were not, but coded so that Lannoo could not tell which was which. After making the radiographs and separating out the frogs with spongiform bone, Lannoo decoded the specimens and found that none of the parasite-infected frogs had spongiform bone. This confirmed his suspicion that spongiform bone was a contraindication of parasitism. "I've looked at 600 frogs now," Lannoo told me in mid-November 1999. "That's twenty-two species from sixty sites in ten states, collected by at least fifteen different field biologists. Spongiform bone is not present in deformities that can be linked to parasites."

Equally convincing was the direct evidence he compiled that argued against trauma once and for all. Lannoo observed the same dorsal pigment disruptions on the skin of truncated stumps that Kathy Converse of the National Wildlife Health Center in Madison had reported. Stan Sessions told me this was not, in his opinion, inconsistent with amputation. A frog limb amputated at an early stage, Sessions said, would initiate a regenerative response in which the terminal end of the truncation would return to an undifferentiated state. Sessions said it would be like the "end of a melting candle." But Sessions could not explain another morphology that turned up at siginificant frequencies at sites where missing limbs were reported: missing pelvic structures.

Lannoo saw these in his radiographs and the Madison people had observed them in dissections. Some frogs with totally missing limbs were also missing part or even all of their pelvis on the same side. In some cases the pelvis was present, but it was not connected to the spine. Lannoo believed a missing pelvis could not possibly be caused by trauma, since such an injury would almost certainly kill the animal or, at a minimum, leave a horrendous scar. Sessions agreed this was true.

Lannoo tested several of these ideas in the summer of 1999 in a series of amputation experiments he performed at Iowa Lakeside lab on leopard frog tadpoles at various stages of development. The removal of the leg at mid-thigh, Lannoo found, could be easily distinguished from what was seen in the field. Lannoo saw no evidence that trauma caused changes in skin pattern. Trauma did not create spongiform bone segments. Trauma did not cause a regeneration attempt, or anything that resembled one. Trauma did not affect pelvic structure. What trauma did do was leave a scar that was easily observed.

"What this all says," Lannoo told me, "is that whatever is going on in Minnesota, it ain't traumatic amputation."

That same summer, Dave Gardiner and Bruce Blumberg—who'd become a professor at UC Irvine—were awarded a grant of $1.3 million from the EPA to assess the role of "endocrine disruptors that activate retinoid signaling pathways" as a probable cause of deformities in frogs. The basis for their research, which would be carried out at sites in Minnesota and Canada for starters, was their determination that bony triangles are a possible signature of retinoid exposure. When I asked Gardiner if the parasite findings coming out of Minnesota in 1999 influenced their work plan in any way, he said absolutely not.

"The parasite findings are interesting and potentially relevant," he said. "But they have no impact whatsoever on our intended research. Zero. There is no way I would set aside our own hypotheses based on results from a single study. I've signed off on $1.3 million to look for retinoids and that is what I'm going to do."

Within limits, everyone's speculations on the deformities in Minnesota have thus far been proven right. To say, for example, that parasites cause deformities is a true statement. Just as it would be true to say asbestos causes lung cancer. It's a true statement—but not the whole story. There are other causes of lung cancer and other kinds of cancer, just as there are other causes of deformities in frogs. The question as to what threat these other causes pose to humans remains as problematic today as it did on that drizzly morning in 1995 when Cindy Reinitz's students became frightened by the frogs they held in their hands.

In biology, the truth is often well hidden, often nested deep within outer shells of false appearances. Spurious correlations abound. But the truth is a reality we all live with, whether we perceive it or not. And the truth, after more than four years of investigation, was that nobody could say with any certainty why frogs in Minnesota were deformed or what that meant—except that it was one way among many to do away with an entire class of animals.

Deformities kill frogs—that much is certain. Although adult frogs are much less numerous than newly metamorphosed juveniles—and also widely dispersed and not as easy to catch—nobody ever found any significant deformities in adult animals. So it is presumed that the limb abnormalities are incompatible with survival. In 1998, Dave Hoppe found large numbers of frogs dying at CWB in midsummer, right at metamorphic climax.

Dave Gardiner, Sue Bryant, and Bruce Blumberg were at CWB one day in July 1998 and saw the effects of the die-off firsthand. Frog corpses choked the weedline along the shore. Gardiner and Bryant collected a large number of the less-decomposed specimens to take back to Irvine for examination. And if all the dead frogs were an indication, it wasn't just because limb defects impeded locomotion. Rather, it seemed that limb abnormalities might just be one effect among a whole range of responses that resulted from some environmental insult—the other notable response being death. Pieter Johnson lamented to me that he regretted not being able to see CWB just once, just to know what it looked like. "I have this image," he said, "of smoke rising from the waters."

But whether the deformities themselves would thus have an impact on the overall abundance of frogs was uncertain. As David Merrell observed in his natural history of the leopard frog, a population of frogs is maintained at a more or less constant level whenever just two offspring from each female make it successfully to reproductive maturity. Biologists call additions to the breeding pool "recruitment." Frog populations fluctuate, often depending on prevailing weather conditions and other variable factors, but over time recruitment is usually a balancing act. Most frog populations do not disappear entirely in bad times, nor do they grow without limit under favorable conditions. Obviously, only a very small percentage of total offspring are successfully recruited, since a female leopard frog lays as many as five thousand eggs. The inescapable arithmetic tells us that so few juveniles survive in the first place that the effect of anything that adds to the mortality rate may be all but undetectable. Death is a way of life for frogs. A limb deformity may be just one more proximate cause of death in a species where the odds of dying before your time are so close to a hundred percent that it scarcely matters how it happens.

When I asked Mike Lannoo if it wasn't inevitable that high rates of deformities—say, on the order of 50 percent among juvenile frogs—would reduce the overall number of animals in an affected population, he said he wasn't at all sure that would be the case. Maybe, maybe not, he told me. Lannoo thought some people would probably argue that, except for seasonal variations, wetlands would tend to have a more or less fixed carrying capacity for frogs and survivorship is a kind of set figure that is roughly constant. Even with a high incidence of deformities, there might still be more than enough healthy frogs to fill nature's quota. Lannoo's own view was that seasonal weather conditions were actually a significant consideration because so many leopard frog populations in the Midwest breed in shallow, ephemeral wetlands that sometimes dry up in midsummer. This creates an endless cycle of boom and bust among the frogs. Lannoo thought that in a very dry year, everything would go to hell. Adding deformities into the mix probably wouldn't have that much of an impact. In better times, he said, he thought it was indeed possible that a boom could be short-circuited by limb abnormalities that would cause recruitment to fall off and the population to decline. "But I really don't know," he said. "And I don't think anybody else does, either."

It *is* tempting to think of frogs as an inherently indestructible class of animals. Their reproductive "strategy" as Dave Hoppe called it, is simple: One pair of frogs gives rise to thousands or even tens of thousands of offspring. Almost all of them die. But there are so many that a few always make it through and the beat goes on. Frogs have been around for hundreds of millions of years, surviving good conditions and bad, outliving other species too numerous to count that have come and gone from the face of the earth in that time, because this strategy works.

Or at least it always used to.

Dave Hoppe's gut feeling that the frog numbers at the lake known as CWB were declining was another ripple in a rising tide of such observations that herpetologists had been talking about and worrying over for the better part of twenty-five years. In the early 1970s, Hoppe's old friend Bob McKinnell had already noticed a marked reduction in the abundance of leopard frogs throughout Minnesota. Other researchers remarked on the decline as well. Not only were leopard frogs becoming scarce relative to their former numbers, they were also smaller and "sicklier." It was getting harder all the time to catch what McKinnell called "lunker" adult leopard frogs—those robust animals whose snout-to-vent length exceeded one hundred millimeters. And people at scientific laboratories and medical schools who bought frogs from Minnesota suppliers complained that the animals were much more often ill or dead on arrival than in the past. In 1975, the Minnesota Department of Natural Resources banned commercial harvesting of larger *Rana pipiens* after surveys in twelve counties showed very low numbers of the frog. A year later, the DNR found that those already depressed populations had declined another 50 percent. McKinnell, who talked regularly with longtime commercial frog pickers knew waning frog numbers were for real and that the declines had been going on for some time.

It was determined at the time that many leopard frogs were succumbing to septicemia—blood poisoning—caused by an aquatic bacterium called *Aeromonas hydrophilia*. But as is often the case in wildlife epidemics, the infection that brings on death was seen as only the immediate cause of the frogs' demise. McKinnell and other researchers thought there had to be another, more comprehensive reason why frogs were suddenly succumbing to *A. hydrophilia* because it was a bacterium routinely present in wetlands and even within the gut of perfectly healthy frogs. In an interview he gave to *Modern Medicine* magazine in 1973, McKinnell speculated that frogs were losing their natural resistance to this common pathogen as a result of "a marked decline in the quality of life." McKinnell said he thought the ultimate cause of the declines was probably an "environmental insult."

"I don't know what that insult is, but a reasonable first guess is insecticides and herbicides," he told the magazine. McKinnell said it was worrisome that the same habitat that would harbor frog populations—generally low-lying wetlands—would also concentrate runoff from agricultural chemicals. He believed that whatever was happening to the environment would eventually do more than merely increase frogs' susceptibility to disease. McKinnell noted that he had begun to observe dead frog-egg masses in the field—something he'd never seen before.

Leopard frogs staged a modest comeback in the upper Midwest—Minnesota reopened these frogs to commercial collecting after 1987—but their numbers never returned to historical levels. Commercial harvesting today yields less than a ton of leopard frogs each year in Minnesota. Before 1973 the annual take was more typically 100,000 pounds. And compared to their ancestors, the frogs remain puny. Dave Hoppe, Bob McKinnell, and Mike Lannoo all told me they rarely see the super-sized, hundred-millimeter-plus adults that were once common in that part of the country.

In a sense, then, it is possible to think of the limb deformities not as a problem in and of themselves, but rather as a lethal symptom of something bigger, something akin to what McKinnell had described two decades before as a "quality-of-life issue." For all its pop culture connotations, the pointy irony of the phrase is perfectly on target. What could be a greater assault on a frog's quality of life than the prospect of death lurking in the places where it makes its home. The world has begun to lose an entire class of animal—among the oldest and best adapted vertebrates on the face of the earth. Frogs are disappearing from places where there was no obvious reason for them to disappear. The proximate cause of death varies. In some cases it is clear. If you build a shopping mall on swampland you kill everything that gets in your way in the process. If you bisect an area of closely knit habitat with an interstate freeway you build a death trap that will do the same work over a longer period of time. But the fear is that not all the proximate causes are nearly so obvious, nor so limited in their effects. What concerns everyone is the disappearance of frogs for no obvious reason. And if frogs are marching away from abundance and into the void, what else is at risk? The same old question endured: What do the places where this is happening have in common besides dead frogs? Environmental unsuitablity is a vague concept, full of meaning yet signifying nothing in particular. But it's what a lot of scientists assume is going on anyway. And gradually the evidence for it has mounted.

In the early 1950s, a group of American Quakers, disillusioned with this country's too-frequent wars, made its way down to Costa Rica—a country with no standing army and, at the time, abundant cheap land for farming. They settled in the lushly forested north-central highlands and foothills of the Corderilla de Tilaran, along the Continental Divide, at a place they named Monteverde. The Green Mountain.

Like most tropical forests, Monteverde is a kind of Garden of Eden. The sheer density of life there is overwhelming. Much of the area is tropical rain forest. At the highest elevations, you enter a rare subcategory: cloud forest. The prevailing trade winds in this part of the world blow from east to west. Pushed in off the Caribbean, laden with moisture, the wind angles up the eastern slopes of the mountains. As it rises, the air cools and its water content condenses, forming an all but perpetual cloud bank that shrouds the upper portion of the jungle at elevations between 4,000 and 6,000 feet. Within the cloud, the forest primeval abounds.

There are more than 2,000 plant species present—about 300 of which are orchids. Orchids, bromeliads, and other epiphytes—plants that grow on other plants—give the jungle here a dense, submarine quality. Within the cloud forest itself conditions are, in a word, wet. Annual rainfall exceeds one hundred inches, but even when it's not raining it seems that it is. Surface condensation and misting from the cloud bank is virtually continuous during all but the driest periods. These primordial conditions favor life, and within the forest live a great variety of animals. There are a hundred different mammals, including the rare jaguar, at least four hundred birds, and more than a hundred species of reptiles and amphibians—though this last category is currently shrinking.

From time immemorial, the most well-known resident of the cloud forest at Monteverde was the brilliantly plumed and very long-tailed resplendent quetzal. The quetzal is a bright-green member of a family of birds called trogons. It has a red breast, and the tail on the showy males grows to as much as eighteen inches in length. The bird was considered sacred by aboriginal people dating far back into pre-Columbian times. More recently it's become a favorite illustration for the letter Q in children's alphabet books. But something happened in the late 1980s that pushed the quetzal into second place among the animals associated with Monteverde.

The Quakers had realized on arriving at Monteverde that the cloud forest was a special place, and they restricted their farming operations to elevations below the main cloud bank. In 1972 they set aside 26,000 acres of the mountaintop as the Monteverde Cloud Forest Biological Preserve, in recognition of the fact that it had become a favorite study area for field biologists. In May of 1964, a second-generation Quaker settler, Jerry Janes, told an American scientist named Jay Savage, who was working in the area, about a brightly colored frog he'd seen right up near the top of the Continental Divide.

The name in Spanish is beautiful: *sapo dorado*. The golden toad. Although it may have been observed from time to time before then, it wasn't until that spring in 1964 that the golden toad became known to science and was given its scientific name, *Bufo periglenes*. Three decades hence it would become the poster frog for worldwide amphibian declines—nothing less than the most famous extinction since the passenger pigeon.

As far as anyone knows, the golden toad never lived anywhere but in a handful of small wetlands atop Monteverde, in the most rarefied part of the jungle. It's a place called the Elfin Forest, where the trade winds top the Continental Divide and cause the trees there to be stunted and bent. The total range of the golden toad encompassed no more than a few square kilometers, but within that small area it was reasonably abundant. In 1987 Martha Crump, a biologist from the University of Florida, began studying the golden toad. Like everyone who ever saw one, Crump was overwhelmed by the animal's fluorescent beauty. In Kathryn Phillips's book *Tracking the Vanishing Frogs*—a crisply written and prescient volume documenting the early days of research into amphibian declines—Crump described the toads as looking like "little jewels" shimmering on the jungle floor.

It's the male that earned the toad its name. Male golden toads were actually an iridescent orange. They were only a couple of inches in length. In the spring, as the rainy season at Monteverde intensified, male golden toads would appear seemingly from nowhere and wait in numbers by their usual breeding sites for the females—which were somewhat larger and darkly mottled—to join them. Crump observed such an assemblage of male golden toads glittering in the mists of the Elfin Forest in the spring of 1987. She was the last person to do so. The following year Crump found fewer than a dozen golden toads; the spring after that she saw only one. Nobody has seen a golden toad since.

Crump informally reported the mystifying disappearance of the golden toad to colleagues at the First World Herpetology Congress in Canterbury, England, in the fall of 1989—a meeting where global amphibian decline became an unstated theme. So many researchers told similar stories of dramatically reduced populations and, in some cases, of being unable to find the very frogs they had been studying for years, that it became apparent to many of the scientists that some unseen changes in the global environment must be responsible for the disappearances. This led to the meeting the following year, in Irvine, California, where the declines were more formally considered, and subsequently to the formation of the Declining Amphibian Populations Task Force. DAPTF's mission to determine the status of amphibians around the world—and to try to understand any hidden causes for their decline—was a tall order. Many species in many places appeared to be less abundant than they had once been. And a few, like the golden toad, had evidently gone completely extinct.

The extinction of the golden toad was consistent with one of the more general observations about amphibian disappearances: the declines were most dramatic in higher elevation habitats, especially tropical ones. The rain forests of eastern Australia, where at least fourteen highland species had been reduced by as much as 90 percent or more, were hard hit. So were many of the mountainous jungles of Central America. But dramatic declines were seen elsewhere—notably in the western United States, where researchers had an advantage they usually didn't elsewhere: strong historical records of amphibian abundance.

In the 1920s, terrestrial vertebrates, including frogs, were extensively surveyed in Yosemite National Park. In the early 1990s, field biologist Gary Fellers from the U.S. Geological Survey in Point Reyes, California, resurveyed the area along the same transect—documenting frog prevalence at the very same habitats. Fellers found a significant reduction in the presence of frogs. Since then Fellers has compiled data from more than 4,000 high-altitude sites in the Sierra Nevada and come to an inescapable conclusion. Frog populations have crashed on the western slopes of the mountains—the side facing California's intensely agricultural Central Valley. Fellers has hypothesized that the cause is aerial deposition of pesticides that are blown up the mountainsides by the prevailing winds. He's found residues of pesticides in ponds where frogs should be living but aren't. Unfortunately, while Fellers's work continues in the mountains, no one is looking at wetlands down in the valley, on the farms. Just as in the Minnesota deformities investigation, pesticides are the prime suspect—but their source remains off-limits to investigators.

For a long time, hard evidence like Fellers's was in short supply. Throughout the mid-1990s, a fierce debate raged among herpetologists over the severity of worldwide amphibian declines. Scientists argued with one another about whether the disappearances were really outside the range of normal population fluctuations. Were the declines in fact occurring simultaneously in geographically diverse areas and thus signaling some cause above and beyond the ongoing decline of biodiversity associated with habitat destruction and modification that resulted from human encroachment?

But the disagreements faded. David Green, of McGill University in Montreal, told me he'd been among the skeptics who'd come to believe the declines were quite real and certainly a cause of concern. Green, who is also the head of Canada's endangered species program, said he was finally convinced of the problem with the complete disappearance of the leopard frog from British Columbia. Leopard frogs, he said, are one of perhaps ten species of frog that are in decline in his country—a situation he finds utterly perplexing. "This is happening in Canada," he said to me, "which for the most part is pristine compared to the United States."

Green's conversion is one of many Dave Wake has seen. "I think we're close to consensus now," Wake told me when I spoke with him on the phone. The question isn't whether many of the world's more than 4,000 amphibian species are disappearing, Wake said, but rather why they are and what that portends for us. The extent of the problem, he said, was probably only just being recognized. Vast areas of amphibian habitat—notably in Mexico and all the African continent—had yet to be extensively surveyed.

Amphibians, Wake added, might not be literal environmental "sentinels," but there is a warning implicit in their demise all the same. "I'm not sure that amphibians as a group are more of a bioindicator than any other class of organisms," Wake said. "In some cases, they're doing well in areas that are obviously environmentally degraded. But they are sending us a message. And it's hard to decipher. Amphibians are bathed in their environment in ways we are not. It's not quite right to think of them as canaries in a coal mine, because this coal mine we're all in is one we can't get out of. This world is all we have. We've got no place to go."

Interior secretary Bruce Babbitt shared Wake's concerns, and the federal response he initiated following the spring briefings that Wake and Dave Gardiner had participated in was eventually named the Taskforce on Amphibian Declines and Deformities—TADD. Babbitt had visited Monteverde and was also familiar with Gary Fellers's studies—the USGS is part of the Interior department—which he was interested in seeing better-funded. "What these scientists have done for me," Babbitt told me in a phone interview, "has been to explain that, for the first time, like a flash of lightning in the night, we're seeing illuminated a landscape of potential extinction that extends all the way around the world."

It was the suddenness of this widespread perception among scientists that seemed to impress Babbitt the most. And this was true with respect to the deformities problem as well. He told me he thought Dave Gardiner's presentation at the first Interior briefing in March 1998 had "driven" his decision to link research into the declines with the investigation into the deformities. "There's a lot of research underway out there now," Babbitt said. "What's needed is to get people together to tell us how to use our resources effectively."

TADD got off to an uneven start in early 1999. An initial funding request for $8 million—money that was to come from other budgets within the Interior department—was reduced in Congress to $3 million. Virtually all of this was to be spent on surveying amphibian populations in the United States, with a good-sized chunk earmarked for Gary Fellers's work in California. The deformities investigation didn't appear to figure anywhere in this for the time being. Meanwhile, many field herpetologists who were already working with the DAPTF couldn't understand why another program was necessary and also saw the multiagency leadership back in Washington as a classicly impotent and stifling bureaucracy. Babbitt had told me that there was agreement among the heads of the various agencies he'd brought into TADD that it should not become a big, unwieldy, "Apollo mission" type of program. For now, it appeared that TADD was neither the centralized, "top-down" organization that Dave Wake and George Rabb had hoped for, nor was it a devouring monster ready to pounce on existing declines and deformities initiatives. It was what it was—a first, tentative official recognition of global problems affecting amphibians. Though the problem seemed to have arisen suddenly, Babbitt thought it had been reasonably considered. "Scientists are cautious," Babbitt said. "They're careful and they're skeptical. They've taken years of arguing amongst themselves to reach some of the conclusions we're now seeing. But there is agreement. And the time to start doing something about it is now."

If the decline problem and the deformities problem did not receive equal treatment from TADD, both were nonetheless much in evidence in central Minnesota, not far from Brainerd, at CWB. During the summer of 1999, following the poor reproduction success and massive mid-summer frog kills of 1998, Dave Hoppe became convinced that CWB was dying. There simply were very, very few frogs to be found in the lake anymore— and most that hung on were even more severely deformed than ever. In addition to the usual leg problems, abnormalities of the mouth increased. Frogs had lower jaws that did not join together at the chin. Incredibly, even the limb defects looked worse. Hoppe increasingly observed a malformation in which a complete, miniature set of hind legs dangled by a thread of skin from one of the primary legs. He called this morphology "satellite limbs" and doubtless would have found a great many of them had he been able to find frogs. But there were next to none all summer long. In July, during a visit Dave Gardiner made to CWB, a four-hour collection effort netted only sixteen mink frogs. All of them were abnormal. It was the largest sample Hoppe managed all summer.

In the spring of 1999, an American biologist from the University of Miami named Alan Pounds solved the mysterious disappearance of the golden toad—along with the simultaneous catastrophic crashes of more than twenty other frog species in the highlands of Costa Rica. Pounds, who had lived and worked at Monteverde for many years and had served as director of the Tropical Science Center at the Cloud Forest Preserve, had already demonstrated in a landmark statistical study in 1997 that these declines were outside the range of any conceivable normal population fluctuation. Three years prior to that, Pounds and Martha Crump, the last person on earth to see a golden toad in the wild, had hypothesized that amphibian declines on tropical mountains might be related to climate change. And that turned out to be the case.

Pounds, working with local meteorological data, stream-flow analyses, and bird demographics compiled by his colleague at Monteverde, Michael Fogden, realized that the series of localized population reversals that began with the 1987 decline and eventual disappearance of the golden toad coincided with the periodic heating of sea-surface temperatures known as El Niño that were superimposed on a longer-term increase in average ocean temperature associated with global warming. These changes at sea were amplified in adjacent tropical highlands and in turn caused the base of the perpetual cloud atop Monteverde to begin to lift. As the lower limits of the cloud moved up the mountainside, warmer and drier conditions moved in behind. Dry spells now lasted longer and occurred at higher elevations. The ecology of the steep mountain forest changed. Under all conditions, life on Monteverde is highly stratified; many species exist only within narrow boundaries of altitude. As the mists lifted and became thinner and less plentiful, animals began to move upward. Those that could not relocate higher declined. Pounds found that, in addition to nearly 40 percent of the local frog species, lizards had also declined dramatically. Two lizard species found only at upper elevations had disappeared entirely by 1996. A third, which thrives in drier conditions, remained stable. Meanwhile, fifteen species of birds formerly found at lower elevations moved up toward the mountaintop. Keel-billed toucans, once found only in the surrounding lowlands, took up residence in the cloud forest, side-by-side with the resplendent quetzal.

These, then, were the famous last days of the golden toad. Small, incremental changes in the local weather pattern brought on by immense, global changes in overall climate raked inexorably over the slopes of Monteverde, changing the equation of life for everything there. Earthbound amphibians, which unlike birds could not fly away from this sort of trouble, instead gave in. For the golden toad, which already lived as high as it could and nowhere else, it was the end. After many millennia of success, the golden toad succumbed to a change to which it could not adapt fast enough to survive. The vanishing point was arrived at.

Pounds's findings were published in a cover article in *Nature*, the world's most prestigious scientific journal, in April 1999, just weeks before Pieter Johnson's and Stan Sessions's papers appeared in *Science*. One reason I'd gone to see Pounds a few months earlier was to answer the question of whether larger environmental factors influence more discrete phenomena like parasitism.

When I visited Pounds in late 1998, he and his wife, Marlene, were living temporarily in a borrowed house next door to what had been their home and laboratory before it was crushed by an enormous tree falling over on it that fall. Pounds is a pleasant, good-humored, and unusually gracious man who reminded me—somewhat jarringly given the locale and the nature of his work—of Steve Martin, the comic actor. Originally from Arkansas, he spoke in measured but assured cadences, a faint twang occasionally disappearing when he would unwittingly shift into Spanish. We talked at length over the course of several days.

Pounds told me that global warming was a sort of umbrella cause for what was undoubtedly a whole catalogue of more proximate causes for amphibian declines—and also an issue that ought to be of major concern as a human health consideration. "We've observed a pattern here and our responsibility is to sound an alarm," he said. He said it probably wasn't warmer temperatures and drier conditions in and of themselves that led to the demise of the golden toad. Rather, he presumed that the climate shift weakened the toads, making them vulnerable to a host of pathogens or parasites that would otherwise not pose a threat. One of these was something called a "chytrid fungus," which was not previously known to infect vertebrates, but which had recently been discovered in Australia and Central America in frog carcasses recovered from massive die-offs. Although it was not clear whether the chytrid fungus was a cause or an effect of amphibian declines, it or something like it was probably the immediate cause of death for the golden toad, though Pounds said nobody would ever know. With extinction, the case is closed forever.

"At the time of the crash we weren't aware of what was happening," Pounds said. "Nobody looked at the animals to see what killed them. The evidence that fungi can be lethal is pretty suggestive so I would think that a fruitful area for future investigation would be possible interactions between climate and epidemics. That's also an area that's under close scrutiny in the human health area. There's a lot of interest in the effects of weather on disease. I wouldn't say we could rule out airborne contaminants in the frog declines at this point. It's still a possibility. And if you look at the broader picture, with the decline and disappearance of lizard populations, this suggests that there's more than one mechanism at work. If we're correct that this is all part of a constellation of changes that related to one another, it seems unlikely that the same pathogen that wiped out amphibians would also be affecting reptiles. So there may be more than one thing going on. If climate is an underlying factor, it could orchestrate a variety of impacts. That's what's so frightening about it. It can affect populations in manifold pathways and all kinds of things could happen."

Andy Blaustein thought Pounds's work was terrific, and another object lesson in interconnectedness in nature. "This is very important," Blaustein told me when I spoke with him about Pounds's findings on climate effects. "It's a convincing scenario for why the golden toad and other species went down the tubes. It also shows how incredibly complex these environmental factors can be."

Mike Lannoo was likewise impressed. He told me how much he admired Pounds's work and his lifestyle. With his track record, Lannoo told me, Pounds could have his pick of plum academic posts in the States. Instead, he chose to remain in the field, scraping by on occasional grants and working harder than he'd have to as a lecturer back home. "I think Alan's become a kind of blue-collar hero to a lot of field biologists," Lannoo said. And the significance of his work, Lannoo added, was undeniable. This, Lannoo maintained, was the first animal extinction attributable to modern climate change—and a harbinger of things to come. "People who say global warming won't be a problem argue that animals will simply shift to more suitable habitats as change occurs," Lannoo said. "Alan's results show there are limits to that."

———————

So the frogs are telling us a lot—so much, perhaps, that we cannot yet fully understand the message. Frogs are succumbing to parasites, to pesticides, to increases in ultraviolet radiation, to global warming. The earth is changing and the frogs are responding.

On a cold day in December of 1998, I went with Mike Lannoo from his home in Muncie down to Athens, Ohio, to Ohio University where he made his radiographs in the lab of an old friend and colleague, Joe Eastman. It was a pleasant trip. The X-ray machine was an odd contraption—like a tall toaster oven—that made radiographs in long, low-intensity exposures. The results were disconcertingly pretty—bone-white images of skeletons in varying degrees of disarray against a black void.

On the drive back to Muncie after dark we talked in the car about evolution, about frogs, about whether the environment was in trouble. Lannoo thought there were many things we could do to make the world cleaner and more habitable—and certainly many of the things that would benefit frogs, like maintaining habitats, were relatively easy. But he said thinking about it made him tired. It had been a very long day. The night sat heavily on the invisible landscape as we rushed through it, passing by a few other cars and the occasional blaring billboard.

"What I think is this," Lannoo said after we'd ridden in silence for a while. "I think there's a lot we can't do anything about. The earth will take care of itself. Time is on its side. We don't have to save the world. We have to save ourselves."

———————

One of the things I'd noticed about the deformities problem was that it always seemed to have a way of starting over again from time to time, in familiar but often inexplicable ways. In the spring of 1999, Dave Hoppe was visited by two researchers from Japan who discussed an outbreak of deformed frogs in a public park near the industrial city of Kitikyushi in their country. Tests over the course of several seasons had ruled out genetic mutations, parasites, and trauma. A contaminant in the water was suspected; work to identify it was ongoing.

The following August, just days after Dan Sutherland confirmed the presence of *Ribeiroia* infections in frogs from CWB, Hoppe got a call from a teacher at a school near the town of Wheaton in west-central Minnesota. The teacher wanted to report deformed frogs in a manmade pond on the school grounds. Hoppe went out to investigate the next day.

Hoppe told me he found a shallow pond, perhaps five feet deep at the most, roughly T-shaped and perhaps fifty feet in length on its long axis. It was about twenty years old. Kids had thrown fish in the pond over the years, and Hoppe found bullheads and fathead minnows living in it. He couldn't find any snails. He did collect fifty-three leopard frogs. About 40 percent were abnormal. Hoppe thought at least half the abnormalities were amputations resulting from trauma. But the rest—extra legs, truncations, skin webbings, and so on—seemed very much like the deformities he was used to seeing at CWB. Hoppe rushed half a dozen specimens down to Dan Sutherland. There were no *Ribeiroia* in them.

In the spring of 1999, Mike Lannoo passed through Minnesota on his way down to the Lakeside Lab in Iowa and stopped off to see me. My wife and I had just bought some land where we were about to begin building a new house and I took Lannoo and his son, Pete, over to see it.

For years I had wondered about the frogs that lived all around us. At our old place we had a large pond, and in the winter, as our kids skated over its frozen surface, I sometimes tried to imagine the frogs beneath them in the frigid darkness, waiting, taking unseen breaths, hearts beating slowly, slowly, resting inside a world that was inside of them. Their destiny was, as it had been for millions of years, to come back in the spring. At our new home, we'd have two ponds.

There was still a hint of spring in the air the day we went to the new place. Lannoo and I walked to the crest of a low hill and surveyed the landscape—a mix of hardwood windbreaks and winding hills interspersed with newly built homes and a few farms that still hung on in the face of people like us coming into the country. Our five acres is high, and Lannoo and I could see clear to downtown Minneapolis twenty-five miles away.

"Mmm," he said. "Are those your wetlands over there?"

We walked down to one of the ponds. Pete bounded ahead, his blond hair level with the tops of the high grasses. As Lannoo and I came down in among the cattails along the shoreline we could hear the plopping of frogs jumping into the water to get out of our way. We stood quietly. Then Lannoo pointed down, to a spot not four feet away. I stared for a minute, then finally saw the angled body and the eyes held just above the surface of the water.

"Green frog," Lannoo said, smiling.

I looked back out over the hills. The sun was high and the air seemed impossibly clear. The breeze felt wonderful, threading softly though the grass where we stood. I wondered how long this had all been in place, what we'd make of it when we lived here. When I looked back down the frog was gone, leaving behind only a ripple where it had been.

Epilogue

SINCE THE FIRST PUBLICATION OF *A PLAGUE OF FROGS* I'VE BEEN repeatedly asked whether "they ever figured out what's causing the deformities." It's a reasonable question. It's *the* question. But the answer remains as unsatisfying as ever. Although much has been learned about the ecology of trematode parasites and their contribution to the widespread "natural" incidence of amphibian malformations, the extent of this phenomenon remains unknown. The same can be said for outbreaks of deformities in which parasites do not play a role. Their prevalence and their causes are still a mystery. The evidence that frog deformities involve multiple phenomena with multiple causes—which was the original premise of the investigation—has grown. Jim Burkhart's assessment that frogs are at the center of a "convergence of environmental misfortune" has emerged as a consensus among most researchers.

If you read between the lines of what we now know, you see a problem as worrisome and as complex and scientifically rich as anyone ever imagined it would be. But it has proven hard to solve and the field has narrowed.

As the months of investigation following the discovery at the Ney pond stretched into years and eventually reached over into a new century, frustration and uncertainty took a toll on a number of the participants. Three of the principal agencies involved—the EPA, the NIEHS, and the MPCA—have abandoned their efforts. The MPCA actually planned to return about sixty thousand dollars of its final legislative appropriation for frog work. Sadly, the long-held conviction within the agency's management that the deformities problem was beyond their capabilities has been borne out. Several investigators outside the agency who have reviewed the MPCA's field data say it is all but worthless. The few general observations that might be gleaned from the agency's six years of work are outweighed by the poor design and execution of a field study that never seemed to find a consistent approach.

Meanwhile, no one has proposed any field study that would encompass a scientifically relevant sample of all land uses and wetland types in Minnesota. Agricultural contamination of frog habitat continues at levels still presumed ordinary and safe, but the actual status of the frogs living in such areas remains anyone's guess.

Pieter Johnson and his colleagues succeeded in a complete identification of the parasite that caused the deformities in his original California studies. They also greatly expanded their research, showing that parasites produce different deformities in different species and that the occurrence of parasite-caused outbreaks appeared to be widespread, especially in the western United States. An ambitious field survey of California, Oregon, Washington, Idaho, and Montana by Johnson and Kevin Lunde showed *Ribeiroia ondatrae* caused deformities in nine species of amphibian at more than fifty historical or newly recorded hot spots. In experimental infections of *Bufo boreas*, the western toad, Johnson again demonstrated that he could induce a range of limb malformations that were dose-dependent and similar to those seen in the wild—though not in the same neatly matched ratios he'd achieved with Pacific treefrogs. Johnson said he presumed toads, given their explosive breeding and brief larval stages, probably escaped heavy infection in the wild but not in the experiment.

Whether Johnson's success discouraged other researchers who were working on different causes is debatable, though it seems likely. At a minimum, everyone working on deformities began routinely looking for parasites at their study sites. When the MPCA discovered a new hot spot near the town of Hibbing in northern Minnesota in 2000, parasites immediately became the presumptive cause when the frogs were found to be heavily infected with *Ribeiroia*.

Johnson and other researchers continued to work on experimental infections of *Ribeiroia* in ranid frogs to determine how the response to the parasites may vary in species like leopard frogs or mink frogs—which are much larger and phenologically different from hylid species like the Pacific treefrog. One important difference appears to be that it takes a much more massive infection to produce comparable limb deformities in ranids. This again raises a question about parasites at sites such as the Ney farm, where *Ribeiroia* has been found only in low numbers and the deformities affect mainly *Rana pipiens*. Johnson believes it is possible that infections were heavier in previous years, when the deformities were also apparently more prevalent. But the MPCA's poor year-to-year data and conflicting parasite analysis from early collections may make it impossible to determine what role parasites play at Ney without extensive further study.

Everyone agrees that parasites could be better evaluated as a cause of malformations in the field if tadpoles were collected and evaluated for infections during limb development. This would help rule out the possibility that at least in some cases parasites might also infect frogs *after* limb development and thus have no involvement in the deformity. Also unknown yet is whether parasites can cause outbreaks consisting of only one type of limb malformation, especially the most common of all the deformities, missing limbs. So far, that has not been demonstrated. Johnson told me he agrees with Stan Sessions that trauma probably accounts for a lot more missing legs than many herpetologists ever believed. But he said he doubted that predation would ever cause much more than perhaps 5 percent of a population to be abnormal. Martin Ouellet told me the same.

Which means that the typical deformities outbreak—such as the ones in Vermont, where the malformations consisted of only missing limbs at frequencies approaching 50 percent and where Sessions could not find trematodes—has yet to be understood.

By mid-2001, Johnson had a number of groundbreaking papers out for review or already in press. After fielding offers from a number of leading aquatic ecology laboratories at prestigious universities, he decided to enter the Ph.D. program at the University of Wisconsin. He plans to continue working on parasites, but not on amphibian deformities.

Stan Sessions, whose views have been in many ways vindicated by Johnson's work and by his own ongoing studies, felt unfairly portrayed in *A Plague of Frogs* and remained bitter at what he saw as a betrayal by his former colleagues in Irvine. In a lengthy e-mail exchange in the spring of 2001, Sessions told me he was more convinced than ever that preconceptions about chemical contaminants and indiscriminant speculation over the deformities that has raged on the Internet for years contributed to a regrettable misallocation of resources and an undermining of scientific discourse. He also indicated that he still believes he was right and Dave Gardiner was wrong on the question of whether retinoids were a plausible cause of the deformities.

Sessions told me he and his students have now completed many experiments involving *Ribeiroia* infections that show the parasite does cause deformities in ranid frogs and that the mechanism is in fact a physical disturbance of cells in the developing limb. He said bony triangles occur in these experiments and also in surgical procedures he has used to rearrange developing legs. He reiterated his claim that ranid frogs sometimes "clear" themselves of cysts after metamorphosis, which, if true, would make parasites a possible cause of deformities even in the absence of any proof of their presence in the affected frogs. I asked Dan Sutherland, the parasitologist in La Crosse, Wisconsin, who has collaborated with Pieter Johnson, about this one more time. Again, Sutherland said emphatically that the idea of a frog completely clearing itself of cysts was absurd.

"Frogs may react to a cyst," he said, "but can they totally eliminate every trace? No way. There is always at least some remnant."

I wrote back to Sessions that I, too, was surprised that many readers and reviewers saw him as the villain of the story, a characterization I neither intended nor anticipated. I felt I'd been accurate and fair in depicting a man who tended to drive many people nuts. For the record, I think most of the participants in this story would recognize the Stan Sessions I described. He's an engaging but sometimes maddening genius who went too far in condemning the views of others, to the point of attempting to undermine research he considered a waste of time and money. But there's another side to Sessions—funny, generous, pleasant to be around—that probably went unnoticed by many whose interactions with him were more formal. It's telling, I think, that in bringing me up to date on his work, Sessions said the thing he was most proud of is that a number of his students had been accepted into top-notch graduate programs.

The search for a link between the deformities and exposure to a retinoid has not been futile. Although progress has been slower than they ever imagined it would be, Dave Gardiner, Bruce Blumberg, and their new postdoctoral associate Felix Grun continue to detect a bioactive retinoid in water from CWB. Using new equipment in the field and in Blumberg's lab at Irvine, they have narrowed the culprit to a handful of organic compounds in one fraction of the assay that has now been run many times. Attempts to purify and identify the target compound have proven time consuming and thus far inconclusive. They believe it is possible that the compound they are looking for is one of the supposedly "inert" carrying ingredients in a pesticide formulation—something that would have likely escaped detection in conventional assays for the active ingredients. Grun continues to visit Minnesota in the summer for additional samples.

Much more progress has been made in demonstrating that retinoids can cause the whole array of reported deformities. Using *Xenopus laevis* and a standard synthetic retinoid, Gardiner and his lab team have succeeded in "building" deformed frogs that look just like the deformed frogs seen in the wild—exactly as they said they would. They've also carefully defined windows of time during limb development when various deformities can be induced. Gardiner told me some of these sensitive periods, when exposure to a retinoid can cause a malformation many weeks later, last only hours.

Gardiner's interpretation of the significance of these findings differs sharply from the conclusions reached by the EPA in related experiments on stage-sensitivity to retinoids that were conducted in Duluth and published in 2000. In that paper, the EPA's lead researcher, Sigmund Degitz, reported that high concentrations of retinoic acid can cause limb deformities in frogs, but only if administered in "pulsed" doses at later larval stages, while chronic exposure to such concentrations in early development is fatal. Sublethal amounts of retinoic acid did not produce abnormal hind limbs in the experiment.

"Under continuous exposure scenarios, embryos would not survive to develop to stages sensitive to chemicals that elicit limb malformation through the retinoid-signaling pathways," Degitz wrote. He went on to speculate that, although a pulsed-dose scenario in the wild is possible, the occurrence of deformities in multiple species that develop at different times argues against it.

This is essentially a restatement of the position the EPA staked out before it did the experiment, and Gardiner still doesn't buy it. He thinks the EPA was determined to find a way to dismiss retinoids as a cause of the deformities and tailored the interpretation of their research to do just that. Gardiner said there are any number of ways you can envision "pulsed" exposures caused by periodic elevations of chemical contaminants in ponds—when it rains, when the farmer sprays pesticides, etc.—and probably many more that nobody has thought of yet.

In fact, there is already overwhelming evidence that many deformities result from pulsed exposures—of parasites. Parasites affect multiple species and can also be lethal, but they are not a "chronic" agent.

Gardiner's most exciting discovery is a hot spot almost literally in his own backyard. In the summer of 2000, Gardiner heard from someone on the youth soccer team he coaches about a kid finding some deformed frogs in a small, temporary pond in the foothills just a half-hour drive from the Irvine campus. Gardiner visited the pond and immediately suspected a familiar story: Pacific treefrogs. He collected some specimens and some water anyway. Then, the following spring, his interest intensified dramatically. The rate of malformation surged, getting as high as 30 percent, and included, as Gardiner put it wryly, the "full Monty" of deformities. Then Grun discovered a retinoid activator when he assayed the water samples. Although the work was still quite preliminary, the compound appears to be close to and perhaps even identical to the one they are working on at CWB. This could indicate, Gardiner cautioned, that the assays were in some way contaminated, and that they would have to check and recheck their procedures. But he said they were using careful controls all along and developing others to eliminate the possibility of a false reading.

At the same time, Gardiner felt compelled to send frogs from the site to Dan Sutherland in Wisconsin for parasite examinations. Given the well-established connection between *Ribeiroia* and deformities among Pacific treefrogs in California, going all the way back to Stan Sessions's first work, Gardiner fully expected Sutherland would find them all infected. He assumed any work he would do on the site would have to take into account that a different, well-understood cause of deformities was already in place.

But Sutherland found no *Ribeiroia* in the frogs.

Sutherland told me his careful dissections of fifteen frogs from the site—more than enough to be sure, he said—turned up not a single encysted trematode. Sutherland seemed unsurprised. Most of the frogs from Minnesota that Sutherland has examined have been free of *Ribeiroia*. This batch from California was intriguing, given the history of parasites and that particular frog species, but in the end it was only more evidence that parasites were not the whole story. "For this parasitologist," Sutherland said, "it's pretty obvious there are multiple causes."

For Gardiner, that much has always been a given and still is.

"In a way, having this site doesn't change anything," Gardiner told me. "I mean, I love this site because it neutralizes some things I've had to deal with now for five years. But I hadn't anticipated our receptor assay would go so slowly. This compound we're looking at may or may not be important. So it's frustrating. Because, remember, retinoids are only our hypothesis. We're testing that, we're not saying retinoids cause the deformities. We're trying to find out if they do. In the meantime, it feels a little like everyone else in this has gone home."

The slowdown in research on frog deformities has made news from the field harder than ever to interpret. Rick Levey at Vermont's Agency for

Natural Resources continued field collections at several study sites in the Lake Champlain area. In early 2002 he told me the malformation rates seemed to have declined sharply from what they'd seen in 1996 and 1997. Levey is convinced that trauma—either predator attacks or cannibalism that may occur in overcrowded wetlands during dry years—probably accounts for some of what has been observed in Vermont. But not for all of it. He said frogs that are missing not only legs but also internal skeletal structures remain hard to explain. He also said Doug Fort was still testing what appear to be teratogenic effects from water and sediment samples from Vermont.

In the summer of 2001, Dan Sutherland and Mike Lannoo, jointly funded by the MPCA and the U.S. Geological Survey, revisited most of the known hot spots and a number of control sites in Minnesota. Lannoo said they made several determinations. One was that all of the sites with malformed frogs are disturbed or in some way "unnatural."

"I think we've all but forgotten what a natural wetland looks like," Lannoo said.

In general, Lannoo found that there was a strong "site" effect—that malformed frogs were not found everywhere. The same was true of *Ribeiroia*, which is present in some wetlands, not in others, and completely absent in certain regions. He said they found plenty evidence that *Ribeiroia* is one factor involved with some deformities in Minnesota. Sites that had both malformations and *Ribeiroia* as well as sites without either were the most common observation. There was also evidence, however, that other causes are involved. At a half-dozen sites, Sutherland and Lannoo found deformed frogs but no *Ribeiroia*. They found a similar number of places where, intriguingly, there were *Ribeiroia* but no deformities. Perhaps the biggest puzzle remained CWB. "It was the only place where we saw dead and dying animals along the shoreline," Lannoo said, "including fish. Even though there is *Ribeiroia* present at CWB I doubt that it's the one cause of everything that is happening there."

David Hoppe has continued to work on CWB, recording breeding calls each spring and making weekly specimen collections through each summer. Deformities still turn up, but at lower frequencies and in less severe forms. Hoppe does not know what to make of this apparent trend—which has been seen by other researchers elsewhere—but his worries over the frog population at CWB have deepened. Over the past several seasons frogs have become harder and harder to find there. Leopard frogs appear to be all but gone from the lake, and mink frog numbers are way down as well.

As always, though, there remained a player to be heard from. I spoke with Martin Ouellet in early 2002, as he was finishing his Ph.D. dissertation, which will report on amphibian deformities, diseases, and population declines in Quebec. He told me he has finished working up his data from a decade in the field and hoped to begin analyzing and publishing his findings in the summer of 2002.

Ouellet said he had been taking a lower profile lately. He was disgusted by the poor quality of much of the deformities work in the United States and soured on what he described as shallow, sensational coverage of that work in the media. The only good news, he said, was that the incompetent investigators appeared to have gotten out. The departure of the MPCA was particularly pleasing. "It's excellent," he said. "It's good for science." Ouellet said he thought Pieter Johnson's work was sound and extremely important, and he also said he'd developed greater respect for Sessions, whose views he was "much more sharing," though not to the point of believing parasites were the one and only true cause of frog malformations. Parasites undoubtedly played a role in amphibian deformities, he said. So did trauma. More so, in both cases, than anyone had ever suspected. But no one can yet say how big a role. He told me that he had been "young and very stubborn" in his original insistence that pesticides were without question the primary cause of the deformities he was seeing in Quebec. Now, he said, "we are stuck with a melting pot of many causes." Ouellet also said he has become increasingly more concerned with the loss of frog habitat as wetlands continue to be destroyed by farming and development. "I'm not as crazy about deformities anymore," he said.

Ouellet said he was still disappointed in U.S. researchers, who he says lack the field data to know what kind of a situation they are really looking at. The situation in Canada, meanwhile, is in many ways unchanged. The same pattern of deformities, the same clear, unmistakable link to agriculture, continued through the summer of 2000—the last season of the Canadian Wildlife Service study. Ouellet told me trematode parasites were "everywhere" in Quebec, in both his study sites and his controls. I asked him anew if he thought they were a primary cause of the deformities. "No," he said. "I believe not. Perhaps, *perhaps*, there are more parasites involved in the extra-legged frogs. But then I have extra-legged frogs with no parasites. And, besides, extra-legged frogs are rare. Parasites have no clear connection with many deformities."

Ouellet told me he planned to continue working on amphibians—

and now reptiles, too—but that his interest was shifting to diseases and population declines and habitat loss, problems he said that dwarf the deformities issue.

In fact, frogs continue to tell us about changes in the world in ways that often seem more momentous than limb deformities and their causes. In the spring of 2001, Joseph Kiesecker, of Penn State, reported finding a link between global warming and amphibian die-offs in the United States that was eerily similar to the connection Alan Pounds had established in Costa Rica. Kiesecker, who had been a graduate student under Andy Blaustein when I visited them in Oregon in 1997, became suspicious of climate change after years of monitoring UV experiments in the high Cascades, where Blaustein was studying declines in Cascades frogs and also *Bufo boreas*, the western toad. Both appeared under duress from UV and from a fungal infection by the water mold *Saprolegnia ferax.* Blaustein had long been investigating the role of increasing UV associated with ozone depletion. Then, one spring, Kiesecker waded into a pond to check a toad breeding site where the water usually rose only slightly above his ankles. This time, it was thigh-deep. Eventually, he pieced together an astonishing sequence of cause and effect that tied declining numbers of western toads to climate change.

Kiesecker discovered that what he at first assumed was an unusually wet spring was, in truth, a *normal* spring. "If I hadn't spent all those years babysitting UV experiments in the mountains I'd have never seen this," he said. The deeper water was due to a return to a more normal pattern of snowfall and early spring rain following years of drier conditions. The reduced runoff feeding into mountain ponds had been caused by warming sea surface temperatures, coupled with a string of intense El Niños in the southern ocean that were felt thousands of miles away as a subtle alteration in the local weather in Oregon's higher elevations. Because the western toad lays its eggs in the same place near the shoreline every year, when the egg masses fall to the bottom of the pond they end up in whatever water depth is there to cover them. Water absorbs UV and provides an important shield against excess UV reaching the eggs. In the dry years, toad eggs had been exposed to as much as twice the normal amount of UV because they were in shallow water. The elevated UV exposure, in turn, retarded embryonic development by a few days. But that delay in hatching was just long enough for *Saprolegnia* to attack. Mold spores that normally land on the surface of the egg masses gained enough time to grow and penetrate the protective jelly coating, gradually strangling the developing toadlets within. Thus, global warming—combined with an unimaginably complex chain reaction of marginal ecological shifts—has been killing toads in Oregon. That's a problem, but more important, Kiesecker's study shows how unpredictable and complicated the effects of large-scale environmental changes are, and how profound the outcome of those changes may be in local ecosystems.

Kiesecker's study was reported in an article that appeared on the cover of the journal *Nature* just a couple of weeks after the Bush administration announced that it was canceling a campaign promise to reduce emissions of the greenhouse gas carbon dioxide and that the United States had no intention of complying with the international Kyoto agreement to combat global warming.

Sources

I N ADDITION TO THE SOURCES LISTED BELOW, I RELIED EXTENSIVELY ON THE following books for information and inspiration: *Biology*, by Neil A. Campbell, Benjamin/Cummings, 4th ed.; Mary C. Dickerson's classic, *The Frog Book*, Dover, 1969 ed.; *Okoboji Wetlands*, by Michael J. Lannoo, University of Iowa Press, 1996; David Quammen's epic *The Song of the Dodo*, Simon & Schuster, 1996; the National Audubon Society's *Field Guide to North American Reptiles and Amphibians*, Knopf, 15th printing, 1997; *Developmental Biology*, by Scott F. Gilbert, Sinauer, 5th ed. 1997; *Our Stolen Future*, by Theo Colborn, Dianne Dumanoski, and John Peterson Myers, Plume, 1997; *Biology of Amphibians*, by William E. Duellman and Linda Trueb, Johns Hopkins University Press, 1994; *Tracking the Vanishing Frogs*, by Kathryn Phillips, Penguin, 1995; *Environmental Physiology of the Amphibians*, edited by Martin E. Feder and Warren W. Burggren, University of Chicago Press, 1992; *Amphibians and Reptiles Native to Minnesota*, by Barney Oldfield and John J. Moriarty, University of Minnesota Press, 1994; and the extraordinary celebration of evolution and ecological diversity, *The Book of Life*, edited by Stephen Jay Gould, W.W. Norton, 1993.

Ankley, Gerald T., and Joseph E. Tietge et al. "Effects of Ultraviolet and Metoprene on Survival an development of *Rana pipiens*." *Environment Toxicology and Chemistry* 17, no. 12 (1998).

Badger, David. *Frogs*. Photographs by John Netherton. Stillwater, MN: Voyageur Press, 1995.

Berger, Lee, et al. "Chytridiomycosis causes amphibian mortality associated population declines in the rain forests of Australia and Central America." *Proceedings of the National Academy of Science* 95 (1998).

Blaustein, Andrew R. et al. "Ambient Ultraviolet Radiation Causes Mortality in Salamander Eggs." *Ecological Applications* 5 (1995).

————. "Chicken Little or Nero's Fiddle? A Perspective on Declining Amphibian Populations." *Herpetelogica* 50 (1994).

————. et al. "DNA Repair Activity and Resistance to Solar UV-B Radiation in Eggs of the Red-legged Frog." *Conservation Biology* 10, no. 5 (1996).

————. et al. "Effects of Ultraviolet Radiation on Amphibians: Field Experiments." *American Zoologist,* 38 (1998): 799–812.

———— and Richard A. Holt et al. "Pathogenic Fungus Contributes to Amphibian Losses in the Pacific Northwest." *Biological Conservation* 67 (1994).

———— et al. "UV repair and resistance to solar UV-B in amphibian eggs: A link to population declines?" *Proceedings of the National Academy of Science* 91 (1994).

———— and David Wake, et al. "Amphibian Declines: Judging Stability, Persistence, and Susceptibility of Populations to Local and Global Extinctions." *Conservation Biology* 8, no. 1 (1994).

Blumberg, Bruce, "An essential role for retinoid signaling in anteroposterior neural specification and neuronal differentiation." *Cell and Developmental Biology,* 8 (1997).

———— et al. "An essential role for retinoid signaling in anteroposterior neural patterning." *Development* 124 (1997).

———— et al. "Multiple retinoid-responsive redeptors in a single cell: Families of retinoid 'X' receptors and retinoic acid receptors in the *Xenopus* egg." *Proceedings of the National Academy of Science* 89 (1992).

———— et al. "Novel retinoic acid receptor ligands in *Xenopus* embryos. *Proceedings of the National Academy of Science* 93 (1996).

———— et al. "Organizer-Specific Homeobox Genes in *Xenopus laevis* Embryos." *Science,* 12 July 1991.

———— and Ronald M. Evans. "Orphan Nuclear Receptors—New Ligands and New Possibilities." *Genes & Development* 12 (1998): 3149–55.

Brown, Michael H. "Love Canal and What It Says About the Poisoning of America." *The Atlantic Monthly,* December 1979.

Bruschelli, Giovanna Migliorini, and Gabriella Rosi. "Polymelia Induced by Vitamin A in Embryos of *Bufo vulgaris*." Institute of General Biology, Perugia University, 1971.

Bryant, Peter J., Susan V. Bryant et al. "Biological Regeneration and Pattern Formation." *Scientific American* 237, no. 1 (1977).

Bryant, Susan V., and D. M. Gardiner. "Retinoic Acid, Local Cell-Cell Interactions, and Pattern Formation in Vertebrate Limbs." *Developmental Biology* 152 (1992).

————, Vernon French, Peter J. Bryant. "Distal Regeneration and Symmetry." *Science* 212 (1981).

———— et al. "Patterning in Limbs: The Resolution of Positional Confrontations." In *Experimental and Theoretical Advances in Biological Pattern Formation*. New York: Plenum Press, 1993.

———— David M. Gardiner, Ken Muneoka. "Limb Development and Regeneration." *American Zoologist* 27 (1987): 675–96.

———— and Ken Muneoka. "Views of limb development and regeneration." *Trends in Genetics* 2, no. 6 (1986).

Burkhart, James G., and Henry S. Gardner. "Non-Mammalian and Environmental Sentinels in Human Health: 'Back to the Future?'" *Human and Ecological Risk Assessment* 3, no. 3 (1997).

———— Judy C. Helgen, and Douglas J. Fort et al. "Induction of Mortality and Malformation in *Xenopus laevis* Embryos by Water Sources Associated with Field Frog Deformities." *Environmental Health Perspectives* 106, no. 12 (1998).

Carson, Rachel. *Silent, Spring*. Boston: Houghton Mifflin, 1994 ed.

Cho, Ken W. Y., and Bruce Blumberg et al. "Cooperation between mesoderm-inducing growth factors and retinoic acide in *Xenopus* axis formation." *Developmental Biology* (1991).

Corn, Paul Stephen. "What We Know and Don't Know about Amphibian Declines in the West." In *Sustainable Ecological Systems: Implementing an Ecological Approach to Land Management*, USDA Technical Report RM-247, May 1994.

Fite, Katherine V., Blaustein, Andrew R. et al. "Evidence of Retinal Light Damage in *Rana cascadae*, a Declining Amphibian Species." *Copeia* 4 (1994).

Fort, Douglas J. "Integrated Ecological Hazard Assessment of Waste Site Soil Extracts Using FETAX and Short-Term Fathead Minnow Teratogenesis Assay." *Environmental Toxicology and Risk Assessment*, American Society of Testing and Materials Symposium (1996).

———— et al. "Development of Short-Term, Whole-Embryo Assays to Evaluate Detrimental Effects on Amphibian Limb Development and Metamorphosis Using *Xenopus laevis*." In press.

———— et al. "Effects of Pond Water, Sediment, and Sediment Extracts from Minnesota and Vermont on Early Development and Metamorphosis in *Xenopus*." In press.

————. "Progress Toward Identifying Causes of Mal-Development Induced in *Xenopus* by Pond Water and Sediment Extracts from Minnesota." In press.

———— et al. "Significance of Experimental Design in Evaluating Ecological Hazards of Sediments/Soils to Amphibian Species." *Environmental Toxicology and Risk Assessment*, American Society of Testing and Materials Symposium (1997).

French, Vernon, Peter J. Bryant, and Susan V. Bryant. "Pattern Regulation in Epimorphic Fields." *Science* 193 (1976).

Futuyma, Douglas J. *Evolutionary Biology*. Sunderland, MA: Sinauer, 1998.

Gardiner, David M., and David M. Hoppe. "Environmentally Induced Limb Malformations in Mink Frogs *(Rana septentrionalis)*." *Journal of Experimental Zoology* 284 (1999): 207–16.

————, Bruce Blumberg et al. "Regulation of *HoxA* expression in developing and regenerating axolotl limbs." *Development* 121, (1995).

————, and Susan V. Bryant. "Molecular mechanisms in the control of limb regeneration: the role of homeobox genes." *International Journal of Developmental Biology* 40 (1996).

———— et al. "Evolution of Vertebrate Limbs: Robust Morphology and Flexible Development." *American Zoology* 38 (1998): 659–71.

Gehring, Walter J. *Master Control Genes in Development and Evolution: The Homeobox Story*. New Haven: Yale University Press, 1998.

Gould, Stephen Jay. *Ontogeny and Phylogeny*. Cambridge: Belknap Press of Harvard, 1977.

Green, David M., *Amphibians in Decline: Canadian Studies of a Global Problem*. Society for the Study of Amphibians and Reptiles, 1997.

Hebard, William B. et al. "Hind Limb Anomalies of a Western Montana Population of the Pacific Tree Frog, *Hyla regilla*." *Copeia*, 3 (1963).

Hoppe, David M. "Chorus Frogs and Their Colors." *Ecology of Reptiles and Amphibians in Minnesota: A Symposium*, 1981.

——— et al. "Developmental Features Influencing Color Polymorphism in Chorus Frogs." *Journal of Herpetology* 18, no. 2 (1984).

———. "History of Minnesota Frog Abnormalities: Do Recent Findings Represent a New Phenomenon?" In press.

———. "Thermal Tolerance in Tadpoles of the Chorus Frog *Pseudacris triseriata*." *Herpetologica* 34 (1978).

———. "Using African Clawed Frogs in the Classroom." *Ecology of Reptiles and Amphibians in Minnesota: A Symposium*, 1981.

——— and Robert G. McKinnell. "Minnesota's Mutant Leopard Frogs." *The Minnesota Volunteer*, November/December 1991.

Johnson, Pieter T.J. and Kevin B. Lunde et al. "The Effect of Trematode Infection on Amphibian Limb Development and Survivorship." *Science* 284 (1999).

Kiesecker, Joseph M., and Andrew R. Blaustein. "Influences of Egg Laying Behavior on Pathogenic Infection of Amphibian Eggs." *Conservation Biology* 11, no. 1 (1997).

——— and Andrew R. Blaustein. "Synergism between UV-B radiation and a pathogen magnifies amphibian embryo mortality in nature." *Proceedings of the National Academy of Science* 92 (1995).

Kolata, Gina Bari. "Love Canal: False Alarm Caused by Botched Study." *Science* 208 (1980).

Lannoo, M. J. "The evolution of the amphibian lateral line system, and its bearing on amphibian phylogeny." *Zeithscrift für Zoology and Systemie und Evolution-forschung* 26 (1988).

————. "An Altered Amphibian Assemblage: Dickinson County, Iowa, 70 Years after Frank Blanchard's Survey." *American Midlands Naturalist* 131 (1994): 311–19.

———— et al. "Mummification Following Winterkill of Adult Green Frogs (Ranidae: *Rana clamitans).*" *Herpetological Review* 29 (1998).

————. *Status and Conservation of Midwestern Amphibians.* Ames: University of Iowa Press, 1998.

Larson, Kjersten L., and Michael J. Lannoo et al. "Paedocannibal Morph Barred Tiger Salamanders Ambystoma tigrinium mavortium from Eastern South Dakota." *American Midlands Naturalist* 141 (1999).

Lowcock, L. A., M. Ouellet, and J. Rodrigue et al. "Flow cytometric assay for in vivo genotoxic effects of pesticides in green frogs (*Rana clamitans*)." *Aquatic Toxicology* 38 (1997).

Maden, M. "The homeotic transformation of tails into limbs in *Rana temporarita* by retinoids." *Developmental Biology* 159 (1993).

Materna, Elizabeth J., et al. "Effects of the Synthetic Pyrethroid Insecticide, Esfenvalerate, on Larval Leopard Frogs (*Rana spp.).*" *Environmental Toxicology and Chemistry* 14, no. 4 (1995).

McKinnell, Robert G. "Reduced Prevalence of the Lucke Renal Adenocarcinoma in Populations of *Rana pipiens* in Minnesota." *Journal of the National Cancer Institute* 65 (1979).

————. "Where have all the frogs gone?" *Modern Medicine* 12, November 1973.

Merrell, David. J. "Life History of the Leopard Frog, *Rana pipiens,* in Minnesota." *Bell Museum of Natural History Occasional Papers,* Number 15, 1977.

————. "Natural Selection in a Leopard Frog Population." *Journal of the Minnesota Academy of Science* 35, no. 2 (1969).

Morgan, Marsha K., et al. "Teratogenic Potential of Atrazine and 2, 4-D Using FETAX." *Journal of Toxicology and Environmental Health* 48 (1996): 151–68.

Moriarty, John J., and Delvin Jones. "Minnesota's Amphibians and Reptiles, Their Conservation and Status: Proceedings of a Symposium." *Minnesota Herpetological Society*, 1997.

Mullen, Lina M., and Susan V. Bryant et al. "Nerve dependency of regeneration: The role of Distal-less and FGF signaling in amphibian limb regeneration." *Development* 122 (1996).

Muneoka, Ken, and Susan V. Bryant. "Evidence that patterning mechanisms in developing and regenerating limbs are the same." *Nature* 298, no. 5872 (1982).

————, Susan V. Bryant, and David M. Gardiner. "Growth Control in Limb Regeneration." *Developmental Biology of the Axolotl,* Oxford University Press (1989).

———— et al. "Cellular Contribution from Dermis and Cartilage to the Regenerating Limb Blastema in Axolotls." *Developmental Biology* 116 (1986).

———— et al. "Intrinsic Control of Regenerative Loss in *Xenopus laevis* Limbs." *Journal of Experimental Zoology* 240 (1986): 47–54.

———— et al. "Pattern discontinuity, polarity and directional intercalation in axolotl limbs." *J. Embryol. Exp. Morph.* 93 (1986).

Ohsugi, Kojune, David M. Gardiner, and Susan V. Bryant. "Cell Cycle Length Affects Gene Expresionand Pattern Formation in Limbs." *Developmental Biology* 189 (1997).

Ouellet, Martin. "Amphibian Deformities: Current State of Knowledge." In *Ecotoxicologoy of Amphibians and Reptiles,* SETAC, 1999.

———— et al. "Hindlimb Deformities (Ectromelia, Ectrodactyly) in Free-Living Anurans from Agricultural Habitats." *Journal of Wildlife Diseases* 33 (1997).

Pounds, J. Alan, et al. "Biological Response to Climate Change in a Highland Tropical Forest." *Nature*, 398 (1998).

————. "Tests of Null Models for Amphibian Declines on a Tropical Mountain." *Conservation Biology* 11 (1997).

Pechmann, Joseph H. K., and Henry M. Wilbur. "Putting Declining Amphibian Populations in Perspective: Natural Fluctuations and Human Impacts." *Herpetologica* 50 (1994).

Platt, John R. "Strong Inference." *Science*, 16 October 1964.

Riddle, Robert D., and Clifford J. Tabin. "How Limbs Develop." *Scientific American*, February 1999.

Rudwick, Martin J. S. *Scenes from Deep Time: Early Pictorial Representations of the Prehistoric World*. Chicago: University of Chicago Press, 1992.

Schmidt, Charles W. "Amphibian Deformities Continue to Puzzle Researchers." *Environmental Science & Technology/News* 31, no. 7 (1997).

Sessions, Stanley K. "Morphological Clues from Multilegged Frogs: Are Retinoids to Blame?" *Science* 284 (1999).

———— and Susan V. Bryant. "Evidence That Regenerative Ability Is an Intrinsic Property of Limb Cells in *Xenopus*." *Journal of Experimental Zoology* 247 (1988): 39–44.

————, David M. Gardiner, and Susan V. Bryant. "Compatible Limb Patterning Mechanisms in Urodeles and Anurans." *Developmental Biology* 131 (1989).

————, and Stephen B. Ruth. "Explanation for Naturally Occurring Supernumerary Limbs in Amphibians." *Journal of Experimental Zoology* 254 (1990): 38–47.

Shaw, Margery W. "Love Canal Chromosome Study." Letters, *Science* 209 (1980).

Stebbins, Robert C., and Nathan W. Cohen. *A Natural History of Amphibians*. Princeton: Princeton University Press, 1995.

Stock, Gregory B., and Susan V. Bryant. "Studies of Digit Regeneration and Their Implications for Theories of Development and Evolution of Vertebrate Limbs." *Journal of Experimental Zoology* 216 (1981): 423–33.

Torok, Maureen A., and David M. Gardiner. "Expression of *HoxD* Genes in Developing and Regenerating Axolotl Limbs." *Developmental Biology* 200 (1998).

Van Valen, Leigh. "A Natural Model for the Origin of Some Higher Taxa." *Journal of Herpetology* 8, no. 2 (1974).

Volpe, E. Peter, and Peter A. Rosenbaum. *Understanding Evolution*, 6th ed., New York: McGraw Hill, 1999.

Wanek, N., and D. M. Gardiner et al. "Conversion by retinoic acide of anterior cells to ZPA cells in the chick wing bud." *Nature* 350, no. 6313 (1991).

Wilson, E. O. *Biodiversity*. Washington, DC: National Academy Press, 1988.

Wright, Albert Hazen, and Anna Allen Wright. *Handbook of Frogs and Toads of the United States and Canada*. Ithaca, NY: Comstock, 1995.

William Souder is an award-winning journalist who has written for some of the nation's largest newspapers. He covered the story of Minnesota's deformed frogs for the *Washington Post*. He lives in Stillwater, Minnesota.